"Beautifully lucid . . . Gleick has a novelist's touch for describing his scientists and their settings, an eye for the apt analogy, and a sense of the dramatic and the poetic."
—*San Francisco Chronicle*

"I was caught up and swept along by the flow of this astonishing chronicle of scientific thought. It has been a long, long time since I finished a book and immediately started reading it all over again for sheer pleasure."
—Lewis Thomas, author of *Lives of a Cell*

"*Chaos* is a book that deserves to be read, for it chronicles the birth of a new scientific technique that may someday be important."
—*The Nation*

"Gleick's *Chaos* is not only enthralling and precise, but full of beautifully strange and strangely beautiful ideas."
—Douglas Hofstadter, author of *Gödel, Escher, Bach*

"Admirably portrays the cutting edge of thought."
—*Los Angeles Times*

"A stunning work, a deeply exciting subject in the hands of a first-rate science writer. The implications of the research James Gleick sets forth are breathtaking."
—Barry Lopez, author of *Arctic Dreams*

"An ambitious and largely successful popular science book that deserves wide readership."
—*Chicago Sun-Times*

"There is a teleological grandeur about this new math that gives the imagination wings."
—*Vogue*

"A splendid introduction. Not only does it explain accurately and skillfully the fundamentals of chaos theory, but it also sketches the theory's colorful history, with entertaining anecdotes about its pioneers and provocative asides about the philosophy of science and mathematics."
—*The Boston Sunday Globe*

Also by James Gleick

Nature's Chaos (with Eliot Porter)

Genius: The Life and Science of Richard Feynman

Faster: The Acceleration of Just About Everything

*What Just Happened: A Chronicle from
the Information Frontier*

Isaac Newton

James Gleick was born in New York City in 1954. He worked for ten years as an editor and reporter for *The New York Times,* founded an early Internet portal, the Pipeline, and has written six books: *Chaos, Nature's Chaos, Genius, Faster, What Just Happened,* and *Isaac Newton.* He lives in Key West and New York with his wife.

James Gleick's Web site is www.around.com.

CHAOS

Making a New Science

James Gleick

PENGUIN BOOKS

PENGUIN BOOKS
Published by the Penguin Group
Penguin Group (USA) Inc., 375 Hudson Street, New York, New York 10014, U.S.A.
Penguin Group (Canada), 90 Eglinton Avenue East, Suite 700, Toronto,
Ontario, Canada M4P 2Y3 (a division of Pearson Penguin Canada Inc.)
Penguin Books Ltd, 80 Strand, London WC2R 0RL, England
Penguin Ireland, 25 St Stephen's Green, Dublin 2, Ireland (a division of Penguin Books Ltd)
Penguin Group (Australia), 250 Camberwell Road, Camberwell,
Victoria 3124, Australia (a division of Pearson Australia Group Pty Ltd)
Penguin Books India Pvt Ltd, 11 Community Centre,
Panchsheel Park, New Delhi – 110 017, India
Penguin Group (NZ), 67 Apollo Drive, Rosedale, North Shore 0632, New Zealand
(a division of Pearson New Zealand Ltd)
Penguin Books (South Africa) (Pty) Ltd, 24 Sturdee Avenue,
Rosebank, Johannesburg 2196, South Africa

Penguin Books Ltd, Registered Offices:
80 Strand, London WC2R 0RL, England

First published in the United States of America by Viking Penguin Inc. 1987
Published in Penguin Books 1988
This edition with a new afterword by James Gleick published 2008

3 5 7 9 10 8 6 4 2

Portions of this book first appeared, in different form, in *The New York Times Magazine*,
as "Solving the Mathematical Riddle of Chaos" and "The Man Who Reshaped Geometry."

Pages 361–362 constitute an extension of this copyright page.

LIBRARY OF CONGRESS CATALOGING IN PUBLICATION DATA
Gleick, James.
Chaos: making a new science/James Gleick.
p. cm.
ISBN 978-0-14-311345-4
1. Chaotic behavior in systems. I. Title.
[Q172.5.C45G54 1988]
003—dc19   88—17448

Printed in the United States of America
Set in Melior

To Cynthia

human was the music,
natural was the static . . .

—JOHN UPDIKE

Contents

Strange Attractors   119

A problem for God. Transitions in the laboratory. Rotating cylinders and a turning point. David Ruelle's idea for turbulence. Loops in phase space. Mille-feuilles and sausage. An astronomer's mapping. "Fireworks or galaxies."

Universality   155

A new start at Los Alamos. The renormalization group. Decoding color. The rise of numerical experimentation. Mitchell Feigenbaum's breakthrough. A universal theory. The rejection letters. Meeting in Como. Clouds and paintings.

The Experimenter   189

Helium in a Small Box. "Insolid billowing of the solid." Flow and form in nature. Albert Libchaber's delicate triumph. Experiment joins theory. From one dimension to many.

Images of Chaos   213

The complex plane. Surprise in Newton's method. The Mandelbrot set: sprouts and tendrils. Art and commerce meet science. Fractal basin boundaries. The chaos game.

The Dynamical Systems Collective   241

Santa Cruz and the sixties. The analog computer. Was this science? "A long-range vision." Measuring unpredictability. Information theory. From microscale to macroscale. The dripping faucet. Audiovisual aids. An era ends.

Inner Rhythms   273

A misunderstanding about models. The complex body. The dynamical heart. Resetting the biological clock. Fatal arrhythmia. Chick embryos and abnormal beats. Chaos as health.

Chaos and Beyond 301

New beliefs, new definitions. The Second Law, the snowflake puzzle, and loaded dice. Opportunity and necessity.

CHAOS

Prologue

THE POLICE IN THE SMALL TOWN of Los Alamos, New Mexico, worried briefly in 1974 about a man seen prowling in the dark, night after night, the red glow of his cigarette floating along the back streets. He would pace for hours, heading nowhere in the starlight that hammers down through the thin air of the mesas. The police were not the only ones to wonder. At the national laboratory some physicists had learned that their newest colleague was experimenting with twenty-six-hour days, which meant that his waking schedule would slowly roll in and out of phase with theirs. This bordered on strange, even for the Theoretical Division.

In the three decades since J. Robert Oppenheimer chose this unworldly New Mexico landscape for the atomic bomb project, Los Alamos National Laboratory had spread across an expanse of desolate plateau, bringing particle accelerators and gas lasers and chemical plants, thousands of scientists and administrators and technicians, as well as one of the world's greatest concentrations of supercomputers. Some of the older scientists remembered the wooden buildings rising hastily out of the rimrock in the 1940s, but to most of the Los Alamos staff, young men and women in college-style corduroys and work shirts, the first bombmakers were just ghosts. The laboratory's locus of purest thought was the Theoretical Division, known as T division, just as computing was C division and weapons was X division. More than a hundred physicists and mathematicians worked in T division, well paid and free of academic pressures to teach and publish. These scientists

1

had experience with brilliance and with eccentricity. They were hard to surprise.

But Mitchell Feigenbaum was an unusual case. He had exactly one published article to his name, and he was working on nothing that seemed to have any particular promise. His hair was a ragged mane, sweeping back from his wide brow in the style of busts of German composers. His eyes were sudden and passionate. When he spoke, always rapidly, he tended to drop articles and pronouns in a vaguely middle European way, even though he was a native of Brooklyn. When he worked, he worked obsessively. When he could not work, he walked and thought, day or night, and night was best of all. The twenty-four-hour day seemed too constraining. Nevertheless, his experiment in personal quasiperiodicity came to an end when he decided he could no longer bear waking to the setting sun, as had to happen every few days.

At the age of twenty-nine he had already become a savant among the savants, an ad hoc consultant whom scientists would go to see about any especially intractable problem, when they could find him. One evening he arrived at work just as the director of the laboratory, Harold Agnew, was leaving. Agnew was a powerful figure, one of the original Oppenheimer apprentices. He had flown over Hiroshima on an instrument plane that accompanied the Enola Gay, photographing the delivery of the laboratory's first product.

"I understand you're real smart," Agnew said to Feigenbaum. "If you're so smart, why don't you just solve laser fusion?"

Even Feigenbaum's friends were wondering whether he was ever going to produce any work of his own. As willing as he was to do impromptu magic with their questions, he did not seem interested in devoting his own research to any problem that might pay off. He thought about turbulence in liquids and gases. He thought about time—did it glide smoothly forward or hop discretely like a sequence of cosmic motion-picture frames? He thought about the eye's ability to see consistent colors and forms in a universe that physicists knew to be a shifting quantum kaleidoscope. He thought about clouds, watching them from airplane windows (until, in 1975, his scientific travel privileges were officially suspended on grounds of overuse) or from the hiking trails above the laboratory.

In the mountain towns of the West, clouds barely resemble the sooty indeterminate low-flying hazes that fill the Eastern air. At Los Alamos, in the lee of a great volcanic caldera, the clouds spill across the sky, in random formation, yes, but also not-random, standing in uniform spikes or rolling in regularly furrowed patterns like brain matter. On a stormy afternoon, when the sky shimmers and trembles with the electricity to come, the clouds stand out from thirty miles away, filtering the light and reflecting it, until the whole sky starts to seem like a spectacle staged as a subtle reproach to physicists. Clouds represented a side of nature that the mainstream of physics had passed by, a side that was at once fuzzy and detailed, structured and unpredictable. Feigenbaum thought about such things, quietly and unproductively.

To a physicist, creating laser fusion was a legitimate problem; puzzling out the spin and color and flavor of small particles was a legitimate problem; dating the origin of the universe was a legitimate problem. Understanding clouds was a problem for a meteorologist. Like other physicists, Feigenbaum used an understated, tough-guy vocabulary to rate such problems. *Such a thing is obvious*, he might say, meaning that a result could be understood by any skilled physicist after appropriate contemplation and calculation. *Not obvious* described work that commanded respect and Nobel prizes. For the hardest problems, the problems that would not give way without long looks into the universe's bowels, physicists reserved words like *deep*. In 1974, though few of his colleagues knew it, Feigenbaum was working on a problem that was deep: chaos.

WHERE CHAOS BEGINS, classical science stops. For as long as the world has had physicists inquiring into the laws of nature, it has suffered a special ignorance about disorder in the atmosphere, in the turbulent sea, in the fluctuations of wildlife populations, in the oscillations of the heart and the brain. The irregular side of nature, the discontinuous and erratic side—these have been puzzles to science, or worse, monstrosities.

But in the 1970s a few scientists in the United States and Europe began to find a way through disorder. They were mathematicians, physicists, biologists, chemists, all seeking connections between different kinds of irregularity. Physiologists found a surprising order in the chaos that develops in the human heart, the prime cause of sudden, unexplained death. Ecologists explored the rise and fall of gypsy moth populations. Economists dug out old stock price data and tried a new kind of analysis. The insights that emerged led directly into the natural world—the shapes of clouds, the paths of lightning, the microscopic intertwining of blood vessels, the galactic clustering of stars.

When Mitchell Feigenbaum began thinking about chaos at Los Alamos, he was one of a handful of scattered scientists, mostly unknown to one another. A mathematician in Berkeley, California, had formed a small group dedicated to creating a new study of "dynamical systems." A population biologist at Princeton University was about to publish an impassioned plea that all scientists should look at the surprisingly complex behavior lurking in some simple models. A geometer working for IBM was looking for a new word to describe a family of shapes—jagged, tangled, splintered, twisted, fractured—that he considered an organizing principle in nature. A French mathematical physicist had just made the disputatious claim that turbulence in fluids might have something to do with a bizarre, infinitely tangled abstraction that he called a strange attractor.

A decade later, chaos has become a shorthand name for a fast-growing movement that is reshaping the fabric of the scientific establishment. Chaos conferences and chaos journals abound. Government program managers in charge of research money for the military, the Central Intelligence Agency, and the Department of Energy have put ever greater sums into chaos research and set up special bureaucracies to handle the financing. At every major university and every major corporate research center, some theorists ally themselves first with chaos and only second with their nominal specialties. At Los Alamos, a Center for Nonlinear Studies was established to coordinate work on chaos and related problems; similar institutions have appeared on university campuses across the country.

Chaos has created special techniques of using computers and special kinds of graphic images, pictures that capture a fantastic and delicate structure underlying complexity. The new science has spawned its own language, an elegant shop talk of *fractals* and *bifurcations, intermittencies* and *periodicities, folded-towel diffeomorphisms* and *smooth noodle maps.* These are the new elements of motion, just as, in traditional physics, quarks and gluons are the new elements of matter. To some physicists chaos is a science of process rather than state, of becoming rather than being.

Now that science is looking, chaos seems to be everywhere. A rising column of cigarette smoke breaks into wild swirls. A flag snaps back and forth in the wind. A dripping faucet goes from a steady pattern to a random one. Chaos appears in the behavior of the weather, the behavior of an airplane in flight, the behavior of cars clustering on an expressway, the behavior of oil flowing in underground pipes. No matter what the medium, the behavior obeys the same newly discovered laws. That realization has begun to change the way business executives make decisions about insurance, the way astronomers look at the solar system, the way political theorists talk about the stresses leading to armed conflict.

Chaos breaks across the lines that separate scientific disciplines. Because it is a science of the global nature of systems, it has brought together thinkers from fields that had been widely separated. "Fifteen years ago, science was heading for a crisis of increasing specialization," a Navy official in charge of scientific financing remarked to an audience of mathematicians, biologists, physicists, and medical doctors. "Dramatically, that specialization has reversed because of chaos." Chaos poses problems that defy accepted ways of working in science. It makes strong claims about the universal behavior of complexity. The first chaos theorists, the scientists who set the discipline in motion, shared certain sensibilities. They had an eye for pattern, especially pattern that appeared on different scales at the same time. They had a taste for randomness and complexity, for jagged edges and sudden leaps. Believers in chaos—and they sometimes call themselves believers, or converts, or evangelists—speculate about determinism and free will, about evolution, about the nature of conscious intelligence. They feel that they are turning back a trend in science toward reductionism, the analysis of systems in terms of their constituent parts: quarks, chromosomes, or neurons. They believe that they are looking for the whole.

The most passionate advocates of the new science go so far

as to say that twentieth-century science will be remembered for just three things: relativity, quantum mechanics, and chaos. Chaos, they contend, has become the century's third great revolution in the physical sciences. Like the first two revolutions, chaos cuts away at the tenets of Newton's physics. As one physicist put it: "Relativity eliminated the Newtonian illusion of absolute space and time; quantum theory eliminated the Newtonian dream of a controllable measurement process; and chaos eliminates the Laplacian fantasy of deterministic predictability." Of the three, the revolution in chaos applies to the universe we see and touch, to objects at human scale. Everyday experience and real pictures of the world have become legitimate targets for inquiry. There has long been a feeling, not always expressed openly, that theoretical physics has strayed far from human intuition about the world. Whether this will prove to be fruitful heresy or just plain heresy, no one knows. But some of those who thought physics might be working its way into a corner now look to chaos as a way out.

Within physics itself, the study of chaos emerged from a backwater. The mainstream for most of the twentieth century has been particle physics, exploring the building blocks of matter at higher and higher energies, smaller and smaller scales, shorter and shorter times. Out of particle physics have come theories about the fundamental forces of nature and about the origin of the universe. Yet some young physicists have grown dissatisfied with the direction of the most prestigious of sciences. Progress has begun to seem slow, the naming of new particles futile, the body of theory cluttered. With the coming of chaos, younger scientists believed they were seeing the beginnings of a course change for all of physics. The field had been dominated long enough, they felt, by the glittering abstractions of high-energy particles and quantum mechanics.

The cosmologist Stephen Hawking, occupant of Newton's chair at Cambridge University, spoke for most of physics when he took stock of his science in a 1980 lecture titled "Is the End in Sight for Theoretical Physics?"

"We already know the physical laws that govern everything we experience in everyday life. . . . It is a tribute to how far we have come in theoretical physics that it now takes enormous machines and a great deal of money to perform an experiment whose results we cannot predict."

Yet Hawking recognized that understanding nature's laws on the terms of particle physics left unanswered the question of how to apply those laws to any but the simplest of systems. Predictability is one thing in a cloud chamber where two particles collide at the end of a race around an accelerator. It is something else altogether in the simplest tub of roiling fluid, or in the earth's weather, or in the human brain.

Hawking's physics, efficiently gathering up Nobel Prizes and big money for experiments, has often been called a revolution. At times it seemed within reach of that grail of science, the Grand Unified Theory or "theory of everything." Physics had traced the development of energy and matter in all but the first eyeblink of the universe's history. But was postwar particle physics a revolution? Or was it just the fleshing out of the framework laid down by Einstein, Bohr, and the other fathers of relativity and quantum mechanics? Certainly, the achievements of physics, from the atomic bomb to the transistor, changed the twentieth-century landscape. Yet if anything, the scope of particle physics seemed to have narrowed. Two generations had passed since the field produced a new theoretical idea that changed the way nonspecialists understand the world.

The physics described by Hawking could complete its mission without answering some of the most fundamental questions about nature. How does life begin? What is turbulence? Above all, in a universe ruled by entropy, drawing inexorably toward greater and greater disorder, how does order arise? At the same time, objects of everyday experience like fluids and mechanical systems came to seem so basic and so ordinary that physicists had a natural tendency to assume they were well understood. It was not so.

As the revolution in chaos runs its course, the best physicists find themselves returning without embarrassment to phenomena on a human scale. They study not just galaxies but clouds. They carry out profitable computer research not just on Crays but on Macintoshes. The premier journals print articles on the strange dynamics of a ball bouncing on a table side by side with articles on quantum physics. The simplest systems are now seen to create extraordinarily difficult problems of predictability. Yet order arises spontaneously in those systems—chaos and order together. Only a new kind of science could begin to cross the great gulf between knowledge of what one thing does—one water molecule, one cell of heart tissue, one neuron—and what millions of them do.

Watch two bits of foam flowing side by side at the bottom of a waterfall. What can you guess about how close they were at the top? Nothing. As far as standard physics was concerned, God might just as well have taken all those water molecules under the table and shuffled them personally. Traditionally, when physicists saw complex results, they looked for complex causes. When they saw a random relationship between what goes into a system and what comes out, they assumed that they would have to build randomness into any realistic theory, by artificially adding noise or error. The modern study of chaos began with the creeping realization in the 1960s that quite simple mathematical equations could model systems every bit as violent as a waterfall. Tiny differences in input could quickly become overwhelming differences in output—a phenomenon given the name "sensitive dependence on initial conditions." In weather, for example, this translates into what is only half-jokingly known as the Butterfly Effect—the notion that a butterfly stirring the air today in Peking can transform storm systems next month in New York.

When the explorers of chaos began to think back on the genealogy of their new science, they found many intellectual trails from the past. But one stood out clearly. For the young physicists and mathematicians leading the revolution, a starting point was the Butterfly Effect.

The Butterfly Effect

*Physicists like to think that all you have to do is say,
these are the conditions, now what happens next?*

—RICHARD P. FEYNMAN

THE SUN BEAT DOWN through a sky that had never seen clouds. The winds swept across an earth as smooth as glass. Night never came, and autumn never gave way to winter. It never rained. The simulated weather in Edward Lorenz's new electronic computer changed slowly but certainly, drifting through a permanent dry midday midseason, as if the world had turned into Camelot, or some particularly bland version of southern California.

Outside his window Lorenz could watch real weather, the early-morning fog creeping along the Massachusetts Institute of Technology campus or the low clouds slipping over the rooftops from the Atlantic. Fog and clouds never arose in the model running on his computer. The machine, a Royal McBee, was a thicket of wiring and vacuum tubes that occupied an ungainly portion of Lorenz's office, made a surprising and irritating noise, and broke down every week or so. It had neither the speed nor the memory to manage a realistic simulation of the earth's atmosphere and oceans. Yet Lorenz created a toy weather in 1960 that succeeded in mesmerizing his colleagues. Every minute the machine marked the passing of a day by printing a row of numbers across a page. If you knew how to read the printouts, you would see a prevailing westerly wind swing now to the north, now to the south, now back to the north. Digitized cyclones spun slowly around an ideal-ized globe. As word spread through the department, the other meteorologists would gather around with the graduate students,

11

making bets on what Lorenz's weather would do next. Somehow, nothing ever happened the same way twice.

Lorenz enjoyed weather—by no means a prerequisite for a research meteorologist. He savored its changeability. He appreciated the patterns that come and go in the atmosphere, families of eddies and cyclones, always obeying mathematical rules, yet never repeating themselves. When he looked at clouds, he thought he saw a kind of structure in them. Once he had feared that studying the science of weather would be like prying a jack-in-the-box apart with a screwdriver. Now he wondered whether science would be able to penetrate the magic at all. Weather had a flavor that could not be expressed by talking about averages. *The daily high temperature in Cambridge, Massachusetts, averages 75 degrees in June. The number of rainy days in Riyadh, Saudi Arabia, averages ten a year.* Those were statistics. The essence was the way patterns in the atmosphere changed over time, and that was what Lorenz captured on the Royal McBee.

He was the god of this machine universe, free to choose the laws of nature as he pleased. After a certain amount of undivine trial and error, he chose twelve. They were numerical rules— equations that expressed the relationships between temperature and pressure, between pressure and wind speed. Lorenz understood that he was putting into practice the laws of Newton, appropriate tools for a clockmaker deity who could create a world and set it running for eternity. Thanks to the determinism of physical law, further intervention would then be unnecessary. Those who made such models took for granted that, from present to future, the laws of motion provide a bridge of mathematical certainty. Understand the laws and you understand the universe. That was the philosophy behind modeling weather on a computer.

Indeed, if the eighteenth-century philosophers imagined their creator as a benevolent noninterventionist, content to remain behind the scenes, they might have imagined someone like Lorenz. He was an odd sort of meteorologist. He had the worn face of a Yankee farmer, with surprising bright eyes that made him seem to be laughing whether he was or not. He seldom spoke about himself or his work, but he listened. He often lost himself in a realm of calculation or dreaming that his colleagues found inaccessible. His closest friends felt that Lorenz spent a good deal of his time off in a remote outer space.

As a boy he had been a weather bug, at least to the extent of keeping close tabs on the max-min thermometer recording the days' highs and lows outside his parents' house in West Hartford, Connecticut. But he spent more time inside playing with mathematical puzzle books than watching the thermometer. Sometimes he and his father would work out puzzles together. Once they came upon a particularly difficult problem that turned out to be insoluble. That was acceptable, his father told him: you can always try to solve a problem by proving that no solution exists. Lorenz liked that, as he always liked the purity of mathematics, and when he graduated from Dartmouth College, in 1938, he thought that mathematics was his calling. Circumstance interfered, however, in the form of World War II, which put him to work as a weather forecaster for the Army Air Corps. After the war Lorenz decided to stay with meteorology, investigating the theory of it, pushing the mathematics a little further forward. He made a name for himself by publishing work on orthodox problems, such as the general circulation of the atmosphere. And in the meantime he continued to think about forecasting.

To most serious meteorologists, forecasting was less than science. It was a seat-of-the-pants business performed by technicians who needed some intuitive ability to read the next day's weather in the instruments and the clouds. It was guesswork. At centers like M.I.T., meteorology favored problems that had solutions. Lorenz understood the messiness of weather prediction as well as anyone, having tried it firsthand for the benefit of military pilots, but he harbored an interest in the problem—a mathematical interest.

Not only did meteorologists scorn forecasting, but in the 1960s virtually all serious scientists mistrusted computers. These souped-up calculators hardly seemed like tools for theoretical science. So numerical weather modeling was something of a bastard problem. Yet the time was right for it. Weather forecasting had been waiting two centuries for a machine that could repeat thousands of calculations over and over again by brute force. Only a computer could cash in the Newtonian promise that the world unfolded along a deterministic path, rule-bound like the planets, predictable like eclipses and tides. In theory a computer could let meteorologists do what astronomers had been able to do with pencil and slide rule: reckon the future of their universe from its initial conditions and the physical laws that guide its evolution. The equations describing the motion of air and water were as well known as those describing the motion of planets. Astronomers did not achieve perfection and never would, not in a solar system tugged by the gravities of nine planets, scores of moons and thousands of asteroids, but calculations of planetary motion were so accurate that people forgot they were forecasts. When an astronomer said, "Comet Halley will be back this way in seventy-six years," it seemed like fact, not prophecy. Deterministic numerical forecasting figured accurate courses for spacecraft and missiles. Why not winds and clouds?

Weather was vastly more complicated, but it was governed by the same laws. Perhaps a powerful enough computer could be the supreme intelligence imagined by Laplace, the eighteenth-century philosopher-mathematician who caught the Newtonian fever like no one else: "Such an intelligence," Laplace wrote, "would embrace in the same formula the movements of the greatest bodies of the universe and those of the lightest atom; for it, nothing would be uncertain and the future, as the past, would be present to its eyes." In these days of Einstein's relativity and Heisenberg's uncertainty, Laplace seems almost buffoon-like in his optimism, but much of modern science has pursued his dream. Implicitly, the mission of many twentieth-century scientists—biologists, neurologists, economists—has been to break their universes down into the simplest atoms that will obey scientific rules. In all these sciences, a kind of Newtonian determinism has been brought to bear. The fathers of modern computing always had Laplace in mind, and the history of computing and the history of forecasting were intermingled ever since John von Neumann designed his first machines at the Institute for Advanced Study in Princeton, New Jersey, in the 1950s. Von Neumann recognized that weather modeling could be an ideal task for a computer.

There was always one small compromise, so small that working scientists usually forgot it was there, lurking in a corner of their philosophies like an unpaid bill. Measurements could never be perfect. Scientists marching under Newton's banner actually waved another flag that said something like this: Given an *approximate* knowledge of a system's initial conditions and an understanding of natural law, one can calculate the *approximate* behavior of the system. This assumption lay at the philosophical heart of science. As one theoretician liked to tell his students: "The basic idea of Western science is that you don't have to take into account the falling of a leaf on some planet in another galaxy when you're trying to account for the motion of a billiard ball on a pool table on earth. Very small influences can be neglected. There's a convergence in the way things work, and arbitrarily small influences don't blow up to have arbitrarily large effects." Classically, the belief in approximation and convergence was well justified. It worked. A tiny error in fixing the position of Comet Halley in 1910 would only cause a tiny error in predicting its arrival in 1986, and the error would stay small for millions of years to come. Computers rely on the same assumption in guiding spacecraft: approximately accurate input gives approximately accurate output. Economic forecasters rely on this assumption, though their success is less apparent. So did the pioneers in global weather forecasting.

With his primitive computer, Lorenz had boiled weather down to the barest skeleton. Yet, line by line, the winds and temperatures in Lorenz's printouts seemed to behave in a recognizable earthly way. They matched his cherished intuition about the weather, his sense that it repeated itself, displaying familiar patterns over time, pressure rising and falling, the airstream swinging north and south. He discovered that when a line went from high to low without a bump, a double bump would come next, and he said, "That's the kind of rule a forecaster could use." But the repetitions were never quite exact. There was pattern, with disturbances. An orderly disorder.

To make the patterns plain to see, Lorenz created a primitive kind of graphics. Instead of just printing out the usual lines of digits, he would have the machine print a certain number of blank spaces followed by the letter *a*. He would pick one variable—perhaps the direction of the airstream. Gradually the *a*'s marched down the roll of paper, swinging back and forth in a wavy line, making a long series of hills and valleys that represented the way the west wind would swing north and south across the continent. The orderliness of it, the recognizable cycles coming around again and again but never twice the same way, had a hypnotic fascination. The system seemed slowly to be revealing its secrets to the forecaster's eye.

One day in the winter of 1961, wanting to examine one sequence at greater length, Lorenz took a shortcut. Instead of starting the whole run over, he started midway through. To give the machine its initial conditions, he typed the numbers straight from the earlier printout. Then he walked down the hall to get away from the noise and drink a cup of coffee. When he returned an hour later, he saw something unexpected, something that planted a seed for a new science.

THIS NEW RUN should have exactly duplicated the old. Lorenz had copied the numbers into the machine himself. The program had not changed. Yet as he stared at the new printout, Lorenz saw his weather diverging so rapidly from the pattern of the last run that, within just a few months, all resemblance had disappeared. He looked at one set of numbers, then back at the other. He might as well have chosen two random weathers out of a hat. His first thought was that another vacuum tube had gone bad.

Suddenly he realized the truth. There had been no malfunction. The problem lay in the numbers he had typed. In the computer's memory, six decimal places were stored: .506127. On the printout, to save space, just three appeared: .506. Lorenz had entered the shorter, rounded-off numbers, assuming that the difference—one part in a thousand—was inconsequential.

It was a reasonable assumption. If a weather satellite can read ocean-surface temperature to within one part in a thousand, its operators consider themselves lucky. Lorenz's Royal McBee was implementing the classical program. It used a purely deterministic system of equations. Given a particular starting point, the weather would unfold exactly the same way each time. Given a slightly different starting point, the weather should unfold in a slightly different way. A small numerical error was like a small puff of wind—surely the small puffs faded or canceled each other out before they could change important, large-scale features of the

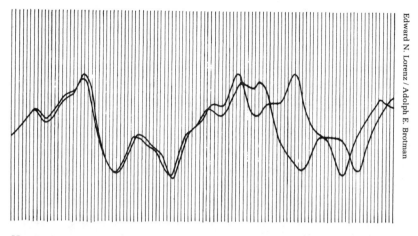

How two weather patterns diverge. From nearly the same starting point, Edward Lorenz saw his computer weather produce patterns that grew farther and farther apart until all resemblance disappeared. (From Lorenz's 1961 printouts.)

weather. Yet in Lorenz's particular system of equations, small errors proved catastrophic.

He decided to look more closely at the way two nearly identical runs of weather flowed apart. He copied one of the wavy lines of output onto a transparency and laid it over the other, to inspect the way it diverged. First, two humps matched detail for detail. Then one line began to lag a hairsbreadth behind. By the time the two runs reached the next hump, they were distinctly out of phase. By the third or fourth hump, all similarity had vanished.

It was only a wobble from a clumsy computer. Lorenz could have assumed something was wrong with his particular machine or his particular model—probably *should* have assumed. It was not as though he had mixed sodium and chlorine and got gold. But for reasons of mathematical intuition that his colleagues would begin to understand only later, Lorenz felt a jolt: something was philosophically out of joint. The practical import could be staggering. Although his equations were gross parodies of the earth's weather, he had a faith that they captured the essence of the real atmosphere. That first day, he decided that long-range weather forecasting must be doomed.

"We certainly hadn't been successful in doing that anyway and now we had an excuse," he said. "I think one of the reasons people thought it would be possible to forecast so far ahead is that there are real physical phenomena for which one can do an excellent job of forecasting, such as eclipses, where the dynamics of the sun, moon, and earth are fairly complicated, and such as oceanic tides. I never used to think of tide forecasts as prediction at all—I used to think of them as statements of fact—but of course, you are predicting. Tides are actually just as complicated as the atmosphere. Both have periodic components—you can predict that next summer will be warmer than this winter. But with weather we take the attitude that we knew *that* already. With tides, it's the predictable part that we're interested in, and the unpredictable part is small, unless there's a storm.

"The average person, seeing that we can predict tides pretty well a few months ahead would say, why can't we do the same thing with the atmosphere, it's just a different fluid system, the laws are about as complicated. But I realized that *any* physical system that behaved nonperiodically would be unpredictable."

THE FIFTIES AND SIXTIES were years of unreal optimism about weather forecasting. Newspapers and magazines were filled with hope for weather science, not just for prediction but for modification and control. Two technologies were maturing together, the digital computer and the space satellite. An international program was being prepared to take advantage of them, the Global Atmosphere Research Program. There was an idea that human society would free itself from weather's turmoil and become its master instead of its victim. Geodesic domes would cover cornfields. Airplanes would seed the clouds. Scientists would learn how to make rain and how to stop it.

The intellectual father of this popular notion was Von Neumann, who built his first computer with the precise intention, among other things, of controlling the weather. He surrounded himself with meteorologists and gave breathtaking talks about his plans to the general physics community. He had a specific mathematical reason for his optimism. He recognized that a complicated dynamical system could have points of instability—critical points where a small push can have large consequences, as with a ball balanced at the top of a hill. With the computer up and running, Von Neumann imagined that scientists would calculate the equations of fluid motion for the next few days. Then a central committee of meteorologists would send up airplanes to lay down smoke screens or seed clouds to push the weather into the desired mode. But Von Neumann had overlooked the possibility of chaos, with instability at *every* point.

By the 1980s a vast and expensive bureaucracy devoted itself to carrying out Von Neumann's mission, or at least the prediction part of it. America's premier forecasters operated out of an unadorned cube of a building in suburban Maryland, near the Washington beltway, with a spy's nest of radar and radio antennas on the roof. Their supercomputer ran a model that resembled Lorenz's only in its fundamental spirit. Where the Royal McBee could carry out sixty multiplications each second, the speed of a Control Data Cyber 205 was measured in megaflops, millions of floating-point operations per second. Where Lorenz had been happy with twelve equations, the modern global model calculated systems of 500,000 equations. The model understood the way moisture moved heat in and out of the air when it condensed and evaporated. The digital winds were shaped by digital mountain ranges. Data poured in hourly from every nation on the globe, from airplanes, satellites, and ships. The National Meteorological Center produced the world's second best forecasts.

The best came out of Reading, England, a small college town an hour's drive from London. The European Centre for Medium Range Weather Forecasts occupied a modest tree-shaded building in a generic United Nations style, modern brick-and-glass architecture, decorated with gifts from many lands. It was built in the heyday of the all-European Common Market spirit, when most of the nations of western Europe decided to pool their talent and resources in the cause of weather prediction. The Europeans attributed their success to their young, rotating staff—no civil service—and their Cray supercomputer, which always seemed to be one model ahead of the American counterpart.

Weather forecasting was the beginning but hardly the end of the business of using computers to model complex systems. The same techniques served many kinds of physical scientists and social scientists hoping to make predictions about everything from the small-scale fluid flows that concerned propeller designers to the vast financial flows that concerned economists. Indeed, by the seventies and eighties, economic forecasting by computer bore a real resemblance to global weather forecasting. The models would churn through complicated, somewhat arbitrary webs of equations, meant to turn measurements of initial conditions—atmospheric pressure or money supply—into a simulation of future trends. The programmers hoped the results were not too grossly distorted by the many unavoidable simplifying assumptions. If a model did anything too obviously bizarre—flooded the Sahara or tripled interest rates—the programmers would revise the equations to bring the output back in line with expectation. In practice, econometric models proved dismally blind to what the future would bring, but many people who should have known better acted as though they believed in the results. Forecasts of economic growth or unemployment were put forward with an implied precision of two or three decimal places. Governments and financial institutions paid for such predictions and acted on them, perhaps out of necessity or for want of anything better. Presumably they knew that such variables as "consumer optimism" were not as nicely measurable as "humidity" and that the perfect differential equations had not yet been written for the movement of politics and fashion. But few realized how fragile was the very process of modeling flows on computers, even when the data was reasonably trustworthy and the laws were purely physical, as in weather forecasting.

Computer modeling had indeed succeeded in changing the weather business from an art to a science. The European Centre's assessments suggested that the world saved billions of dollars each year from predictions that were statistically better than nothing. But beyond two or three days the world's best forecasts were speculative, and beyond six or seven they were worthless.

The Butterfly Effect was the reason. For small pieces of weather—and to a global forecaster, small can mean thunderstorms and blizzards—any prediction deteriorates rapidly. Errors and uncertainties multiply, cascading upward through a chain of turbulent features, from dust devils and squalls up to continent-size eddies that only satellites can see.

The modern weather models work with a grid of points on the order of sixty miles apart, and even so, some starting data has to be guessed, since ground stations and satellites cannot see everywhere. But suppose the earth could be covered with sensors spaced one foot apart, rising at one-foot intervals all the way to the top of the atmosphere. Suppose every sensor gives perfectly accurate readings of temperature, pressure, humidity, and any other quantity a meteorologist would want. Precisely at noon an infinitely powerful computer takes all the data and calculates what will happen at each point at 12:01, then 12:02, then 12:03 . . .

The computer will still be unable to predict whether Princeton, New Jersey, will have sun or rain on a day one month away. At noon the spaces between the sensors will hide fluctuations that the computer will not know about, tiny deviations from the average. By 12:01, those fluctuations will already have created small errors one foot away. Soon the errors will have multiplied to the ten-foot scale, and so on up to the size of the globe.

Even for experienced meteorologists, all this runs against intuition. One of Lorenz's oldest friends was Robert White, a fellow meteorologist at M.I.T. who later became head of the National Oceanic and Atmospheric Administration. Lorenz told him about the Butterfly Effect and what he felt it meant for long-range prediction. White gave Von Neumann's answer. "Prediction, nothing," he said. "This is weather control." His thought was that small modifications, well within human capability, could cause desired large-scale changes.

Lorenz saw it differently. Yes, you could change the weather. You could make it do something different from what it would otherwise have done. But if you did, then you would never *know* what it would otherwise have done. It would be like giving an extra shuffle to an already well-shuffled pack of cards. You know it will change your luck, but you don't know whether for better or worse.

LORENZ'S DISCOVERY WAS AN ACCIDENT, one more in a line stretching back to Archimedes and his bathtub. Lorenz never was the type to shout *Eureka*. Serendipity merely led him to a place he had been all along. He was ready to explore the consequences of his discovery by working out what it must mean for the way science understood flows in all kinds of fluids.

Had he stopped with the Butterfly Effect, an image of predictability giving way to pure randomness, then Lorenz would have produced no more than a piece of very bad news. But Lorenz saw more than randomness embedded in his weather model. He saw a fine geometrical structure, order *masquerading* as randomness. He was a mathematician in meteorologist's clothing, after all, and now he began to lead a double life. He would write papers that were pure meteorology. But he would also write papers that were pure mathematics, with a slightly misleading dose of weather talk as preface. Eventually the prefaces would disappear altogether.

He turned his attention more and more to the mathematics of systems that never found a steady state, systems that almost repeated themselves but never quite succeeded. Everyone knew that the weather was such a system—aperiodic. Nature is full of others: animal populations that rise and fall almost regularly, epidemics that come and go on tantalizingly near-regular schedules. If the weather ever did reach a state exactly like one it had reached before, every gust and cloud the same, then presumably it would repeat itself forever after and the problem of forecasting would become trivial.

Lorenz saw that there must be a link between the unwillingness of the weather to repeat itself and the inability of forecasters to predict it—a link between aperiodicity and unpredictability. It was not easy to find simple equations that would produce the aperiodicity he was seeking. At first his computer tended to lock into repetitive cycles. But Lorenz tried different sorts of minor complications, and he finally succeeded when he put in an equation that varied the amount of heating from east to west, corresponding to the real-world variation between the way the sun warms the east coast of North America, for example, and the way it warms the Atlantic Ocean. The repetition disappeared.

The Butterfly Effect was no accident; it was necessary. Suppose small perturbations remained small, he reasoned, instead of cascading upward through the system. Then when the weather came arbitrarily close to a state it had passed through before, it would *stay* arbitrarily close to the patterns that followed. For practical purposes, the cycles would be predictable—and eventually uninteresting. To produce the rich repertoire of real earthly weather, the beautiful multiplicity of it, you could hardly wish for anything better than a Butterfly Effect.

The Butterfly Effect acquired a technical name: sensitive dependence on initial conditions. And sensitive dependence on initial conditions was not an altogether new notion. It had a place in folklore:

"For want of a nail, the shoe was lost;
For want of a shoe, the horse was lost;
For want of a horse, the rider was lost;
For want of a rider, the battle was lost;
For want of a battle, the kingdom was lost!"

In science as in life, it is well known that a chain of events can have a point of crisis that could magnify small changes. But chaos meant that such points were everywhere. They were pervasive. In systems like the weather, sensitive dependence on initial conditions was an inescapable consequence of the way small scales intertwined with large.

His colleagues were astonished that Lorenz had mimicked both aperiodicity and sensitive dependence on initial conditions in his toy version of the weather: twelve equations, calculated over and over again with ruthless mechanical efficiency. How could such richness, such unpredictability—such chaos—arise from a simple deterministic system?

LORENZ PUT THE WEATHER ASIDE and looked for even simpler ways to produce this complex behavior. He found one in a system of just three equations. They were nonlinear, meaning that they expressed relationships that were not strictly proportional. Linear relationships can be captured with a straight line on a graph. Linear relationships are easy to think about: the more the merrier. Linear equations are solvable, which makes them suitable for textbooks. Linear systems have an important modular virtue: you can take them apart, and put them together again—the pieces add up.

Nonlinear systems generally cannot be solved and cannot be

added together. In fluid systems and mechanical systems, the non-linear terms tend to be the features that people want to leave out when they try to get a good, simple understanding. Friction, for example. Without friction a simple linear equation expresses the amount of energy you need to accelerate a hockey puck. With friction the relationship gets complicated, because the amount of energy changes depending on how fast the puck is already moving. Nonlinearity means that the act of playing the game has a way of changing the rules. You cannot assign a constant importance to friction, because its importance depends on speed. Speed, in turn, depends on friction. That twisted changeability makes nonlinearity hard to calculate, but it also creates rich kinds of behavior that never occur in linear systems. In fluid dynamics, everything boils down to one canonical equation, the Navier-Stokes equation. It is a miracle of brevity, relating a fluid's velocity, pressure, density, and viscosity, but it happens to be nonlinear. So the nature of those relationships often becomes impossible to pin down. Analyzing the behavior of a nonlinear equation like the Navier-Stokes equation is like walking through a maze whose walls rearrange themselves with each step you take. As Von Neumann himself put it: "The character of the equation . . . changes simultaneously in all relevant respects: Both order and degree change. Hence, bad mathematical difficulties must be expected." The world would be a different place—and science would not need chaos—if only the Navier-Stokes equation did not contain the demon of nonlinearity.

A particular kind of fluid motion inspired Lorenz's three equations: the rising of hot gas or liquid, known as convection. In the atmosphere, convection stirs air heated by the sun-baked earth, and shimmering convective waves rise ghost-like above hot tar and radiators. Lorenz was just as happy talking about convection in a cup of hot coffee. As he put it, this was just one of the innumerable hydrodynamical processes in our universe whose future behavior we might wish to predict. How can we calculate how quickly a cup of coffee will cool? If the coffee is just warm, its heat will dissipate without any hydrodynamic motion at all. The coffee remains in a steady state. But if it is hot enough, a convective overturning will bring hot coffee from the bottom of the cup up to the cooler surface. Convection in coffee becomes plainly visible when a little cream is dribbled into the cup. The swirls can be complicated. But the long-term destiny of such a system is obvious. Because the heat dissipates, and because friction slows a moving fluid, the motion must come to an inevitable stop. Lorenz drily told a gathering of scientists, "We might have trouble forecasting the temperature of the coffee one minute in advance, but we should have little difficulty in forecasting it an hour ahead." The equations of motion that govern a cooling cup of coffee must reflect the system's destiny. They must be dissipative. Temperature must head for the temperature of the room, and velocity must head for zero.

Lorenz took a set of equations for convection and stripped it to the bone, throwing out everything that could possibly be extraneous, making it unrealistically simple. Almost nothing remained of the original model, but he did leave the nonlinearity. To the eye of a physicist, the equations looked easy. You would glance at them—many scientists did, in years to come—and say, *I could solve that.*

"Yes," Lorenz said quietly, "there is a tendency to think that when you see them. There are some nonlinear terms in them, but

Adolph E. Brotman

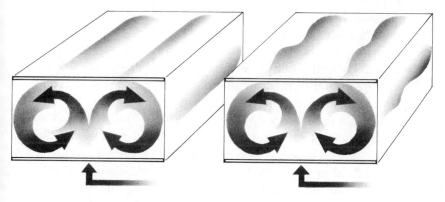

A ROLLING FLUID. When a liquid or gas is heated from below, the fluid tends to organize itself into cylindrical rolls (*left*). Hot fluid rises on one side, loses heat, and descends on the other side—the process of convection. When the heat is turned up further (*right*), an instability sets in, and the rolls develop a wobble that moves back and forth along the length of the cylinders. At even higher temperatures, the flow becomes wild and turbulent.

you think there must be a way to get around them. But you just can't."

The simplest kind of textbook convection takes place in a cell of fluid, a box with a smooth bottom that can be heated and a smooth top that can be cooled. The temperature difference between the hot bottom and the cool top controls the flow. If the difference is small, the system remains still. Heat moves toward the top by conduction, as if through a bar of metal, without overcoming the natural tendency of the fluid to remain at rest. Furthermore, the system is stable. Any random motions that might occur when, say, a graduate student knocks into the apparatus will tend to die out, returning the system to its steady state.

Turn up the heat, though, and a new kind of behavior develops. As the fluid underneath becomes hot, it expands. As it expands, it becomes less dense. As it becomes less dense, it becomes lighter, enough to overcome friction, and it pushes up toward the surface. In a carefully designed box, a cylindrical roll develops, with the hot fluid rising around one side and cool fluid sinking down around the other. Viewed from the side, the motion makes a continuous circle. Out of the laboratory, too, nature often makes its own convection cells. When the sun heats a desert floor, for example, the rolling air can shape shadowy patterns in the clouds above or the sand below.

Turn up the heat even more, and the behavior grows more complex. The rolls begin to wobble. Lorenz's pared-down equations were far too simple to model that sort of complexity. They abstracted just one feature of real-world convection: the circular motion of hot fluid rising up and around like a Ferris wheel. The equations took into account the velocity of that motion and the transfer of heat. Those physical processes interacted. As any given bit of hot fluid rose around the circle, it would come into contact with cooler fluid and so begin to lose heat. If the circle was moving fast enough, the ball of fluid would not lose all its extra heat by the time it reached the top and started swinging down the other side of the roll, so it would actually begin to push back against the momentum of the other hot fluid coming up behind it.

Although the Lorenz system did not fully model convection, it did turn out to have exact analogues in real systems. For example, his equations precisely describe an old-fashioned electri-

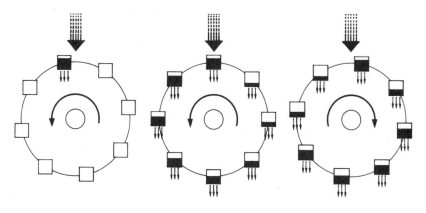

Adolph E. Brotman

THE LORENZIAN WATERWHEEL. The first, famous chaotic system discovered by Edward Lorenz corresponds exactly to a mechanical device: a waterwheel. This simple device proves capable of surprisingly complicated behavior.

The rotation of the waterwheel shares some of the properties of the rotating cylinders of fluid in the process of convection. The waterwheel is like a slice through the cylinder. Both systems are driven steadily— by water or by heat—and both dissipate energy. The fluid loses heat; the buckets lose water. In both systems, the long-term behavior depends on how hard the driving energy is.

Water pours in from the top at a steady rate. If the flow of water in the waterwheel is slow, the top bucket never fills up enough to overcome friction, and the wheel never starts turning. (Similarly, in a fluid, if the heat is too low to overcome viscosity, it will not set fluid in motion.)

If the flow is faster, the weight of the top bucket sets the wheel in motion (*left*). The waterwheel can settle into a rotation that continues at a steady rate (*center*).

But if the flow is faster still (*right*), the spin can become chaotic, because of nonlinear effects built into the system. As buckets pass under the flowing water, how much they fill depends on the speed of spin. If the wheel is spinning rapidly, the buckets have little time to fill up. (Similarly, fluid in a fast-turning convection roll has little time to absorb heat.) Also, if the wheel is spinning rapidly, buckets can start up the other side before they have time to empty. As a result, heavy buckets on the side moving upward can cause the spin to slow down and then reverse.

In fact, Lorenz discovered, over long periods, the spin can reverse itself many times, never settling down to a steady rate and never repeating itself in any predictable pattern.

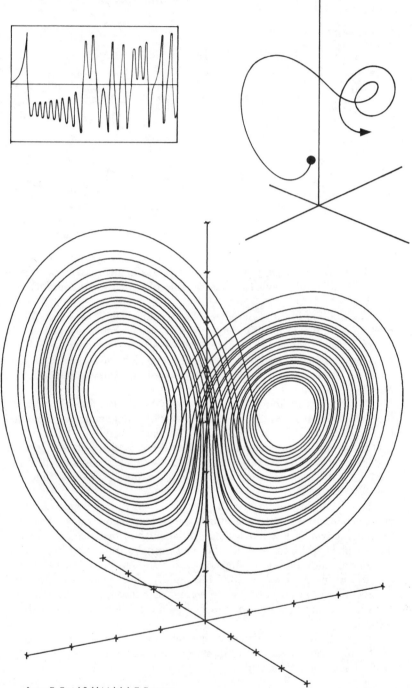

James P. Crutchfield / Adolph E. Brotman

cal dynamo, the ancestor of modern generators, where current flows through a disc that rotates through a magnetic field. Under certain conditions the dynamo can reverse itself. And some scientists, after Lorenz's equations became better known, suggested that the behavior of such a dynamo might provide an explanation for another peculiar reversing phenomenon: the earth's magnetic field. The "geodynamo" is known to have flipped many times during the earth's history, at intervals that seem erratic and inexplicable. Faced with such irregularity, theorists typically look for explanations outside the system, proposing such causes as meteorite strikes. Yet perhaps the geodynamo contains its own chaos.

Another system precisely described by the Lorenz equations is a certain kind of water wheel, a mechanical analogue of the rotating circle of convection. At the top, water drips steadily into containers hanging on the wheel's rim. Each container leaks steadily from a small hole. If the stream of water is slow, the top containers never fill fast enough to overcome friction, but if the stream is faster, the weight starts to turn the wheel. The rotation might become continuous. Or if the stream is so fast that the heavy containers swing all the way around the bottom and start up the other side, the wheel might then slow, stop, and reverse its rotation, turning first one way and then the other.

THE LORENZ ATTRACTOR (on facing page). This magical image, resembling an owl's mask or butterfly's wings, became an emblem for the early explorers of chaos. It revealed the fine structure hidden within a disorderly stream of data. Traditionally, the changing values of any one variable could be displayed in a so-called time series (*top*). To show the changing relationships among three variables required a different technique. At any instant in time, the three variables fix the location of a point in three-dimensional space; as the system changes, the motion of the point represents the continuously changing variables.

Because the system never exactly repeats itself, the trajectory never intersects itself. Instead it loops around and around forever. Motion on the attractor is abstract, but it conveys the flavor of the motion of the real system. For example, the crossover from one wing of the attractor to the other corresponds to a reversal in the direction of spin of the waterwheel or convecting fluid.

A physicist's intuition about such a simple mechanical system—his pre-chaos intuition—tells him that over the long term, if the stream of water never varied, a steady state would evolve. Either the wheel would rotate steadily or it would oscillate steadily back and forth, turning first in one direction and then the other at constant intervals. Lorenz found otherwise.

Three equations, with three variables, completely described the motion of this system. Lorenz's computer printed out the changing values of the three variables: $0-10-0$; $4-12-0$; $9-20-0$; $16-36-2$; $30-66-7$; $54-115-24$; $93-192-74$. The three numbers rose and then fell as imaginary time intervals ticked by, five time steps, a hundred time steps, a thousand.

To make a picture from the data, Lorenz used each set of three numbers as coordinates to specify the location of a point in three-dimensional space. Thus the sequence of numbers produced a sequence of points tracing a continuous path, a record of the system's behavior. Such a path might lead to one place and stop, meaning that the system had settled down to a steady state, where the variables for speed and temperature were no longer changing. Or the path might form a loop, going around and around, meaning that the system had settled into a pattern of behavior that would repeat itself periodically.

Lorenz's system did neither. Instead, the map displayed a kind of infinite complexity. It always stayed within certain bounds, never running off the page but never repeating itself, either. It traced a strange, distinctive shape, a kind of double spiral in three dimensions, like a butterfly with its two wings. The shape signaled pure disorder, since no point or pattern of points ever recurred. Yet it also signaled a new kind of order.

YEARS LATER, PHYSICISTS would give wistful looks when they talked about Lorenz's paper on those equations—"that beautiful marvel of a paper." By then it was talked about as if it were an ancient scroll, preserving secrets of eternity. In the thousands of articles that made up the technical literature of chaos, few were cited more often than "Deterministic Nonperiodic Flow." For years, no single object would inspire more illustrations, even motion pictures, than the mysterious curve depicted at the end, the double spiral that became known as the Lorenz attractor. For the first time, Lorenz's pictures had shown what it meant to say, "This is complicated." All the richness of chaos was there.

At the time, though, few could see it. Lorenz described it to Willem Malkus, a professor of applied mathematics at M.I.T., a gentlemanly scientist with a grand capacity for appreciating the work of colleagues. Malkus laughed and said, "Ed, we *know*—we know very well—that fluid convection doesn't do that at all." The complexity would surely be damped out, Malkus told him, and the system would settle down to steady, regular motion.

"Of course, we completely missed the point," Malkus said a generation later—years after he had built a real Lorenzian water-wheel in his basement laboratory to show nonbelievers. "Ed wasn't thinking in terms of our physics at all. He was thinking in terms of some sort of generalized or abstracted model which exhibited behavior that he intuitively felt was characteristic of some aspects of the external world. He couldn't quite say that to us, though. It's only after the fact that we perceived that he must have held those views."

Few laymen realized how tightly compartmentalized the scientific community had become, a battleship with bulkheads sealed against leaks. Biologists had enough to read without keeping up with the mathematics literature—for that matter, molecular biologists had enough to read without keeping up with population biology. Physicists had better ways to spend their time than sifting through the meteorology journals. Some mathematicians would have been excited to see Lorenz's discovery; within a decade, physicists, astronomers, and biologists were seeking something just like it, and sometimes rediscovering it for themselves. But Lorenz was a meteorologist, and no one thought to look for chaos on page 130 of volume 20 of the *Journal of the Atmospheric Sciences*.

Revolution

Of course, the entire effort is to put oneself
Outside the ordinary range
Of what are called statistics.

—STEPHEN SPENDER

THE HISTORIAN OF SCIENCE Thomas S. Kuhn describes a disturbing experiment conducted by a pair of psychologists in the 1940s. Subjects were given glimpses of playing cards, one at a time, and asked to name them. There was a trick, of course. A few of the cards were freakish: for example, a red six of spades or a black queen of diamonds.

At high speed the subjects sailed smoothly along. Nothing could have been simpler. They didn't see the anomalies at all. Shown a red six of spades, they would sing out either "six of hearts" or "six of spades." But when the cards were displayed for longer intervals, the subjects started to hesitate. They became aware of a problem but were not sure quite what it was. A subject might say that he had seen something odd, like a red border around a black heart.

Eventually, as the pace was slowed even more, most subjects would catch on. They would see the wrong cards and make the mental shift necessary to play the game without error. Not everyone, though. A few suffered a sense of disorientation that brought real pain. "I can't make that suit out, whatever it is," said one. "It didn't even look like a card that time. I don't know what color it is now or whether it's a spade or a heart. I'm not even sure what a spade looks like. My God!"

Professional scientists, given brief, uncertain glimpses of nature's workings, are no less vulnerable to anguish and confusion when they come face to face with incongruity. And incongruity, when it changes the way a scientist sees, makes possible the most important advances. So Kuhn argues, and so the story of chaos suggests.

Kuhn's notions of how scientists work and how revolutions occur drew as much hostility as admiration when he first published them, in 1962, and the controversy has never ended. He pushed a sharp needle into the traditional view that science progresses by the accretion of knowledge, each discovery adding to the last, and that new theories emerge when new experimental facts require them. He deflated the view of science as an orderly process of asking questions and finding their answers. He emphasized a contrast between the bulk of what scientists do, working on legitimate, well-understood problems within their disciplines, and the exceptional, unorthodox work that creates revolutions. Not by accident, he made scientists seem less than perfect rationalists.

In Kuhn's scheme, normal science consists largely of mopping-up operations. Experimentalists carry out modified versions of experiments that have been carried out many times before. Theorists add a brick here, reshape a cornice there, in a wall of theory. It could hardly be otherwise. If all scientists had to begin from the beginning, questioning fundamental assumptions, they would be hard pressed to reach the level of technical sophistication necessary to do useful work. In Benjamin Franklin's time, the handful of scientists trying to understand electricity could choose their own first principles—indeed, had to. One researcher might consider attraction to be the most important electrical effect, thinking of electricity as a sort of "effluvium" emanating from substances. Another might think of electricity as a fluid, conveyed by conducting material. These scientists could speak almost as easily to laymen as to each other, because they had not yet reached a stage where they could take for granted a common, specialized language for the phenomena they were studying. By contrast, a twentieth-century fluid dynamicist could hardly expect to advance knowledge in his field without first adopting a body of terminology and mathematical technique. In return, unconsciously, he would give up much freedom to question the foundations of his science.

Central to Kuhn's ideas is the vision of normal science as solving problems, the kinds of problems that students learn the first time they open their textbooks. Such problems define an accepted style of achievement that carries most scientists through graduate school, through their thesis work, and through the writing of journal articles that makes up the body of academic careers. "Under normal conditions the research scientist is not an innovator but a solver of puzzles, and the puzzles upon which he concentrates are just those which he believes can be both stated and solved within the existing scientific tradition," Kuhn wrote.

Then there are revolutions. A new science arises out of one that has reached a dead end. Often a revolution has an interdisciplinary character—its central discoveries often come from people straying outside the normal bounds of their specialties. The problems that obsess these theorists are not recognized as legitimate lines of inquiry. Thesis proposals are turned down or articles are refused publication. The theorists themselves are not sure whether they would recognize an answer if they saw one. They accept risk to their careers. A few freethinkers working alone, unable to explain where they are heading, afraid even to tell their colleagues what they are doing—that romantic image lies at the heart of Kuhn's scheme, and it has occurred in real life, time and time again, in the exploration of chaos.

Every scientist who turned to chaos early had a story to tell of discouragement or open hostility. Graduate students were warned that their careers could be jeopardized if they wrote theses in an untested discipline, in which their advisors had no expertise. A particle physicist, hearing about this new mathematics, might begin playing with it on his own, thinking it was a beautiful thing, both beautiful and hard—but would feel that he could never tell his colleagues about it. Older professors felt they were suffering a kind of midlife crisis, gambling on a line of research that many colleagues were likely to misunderstand or resent. But they also felt an intellectual excitement that comes with the truly new. Even outsiders felt it, those who were attuned to it. To Freeman Dyson at the Institute for Advanced Study, the news of chaos came "like an electric shock" in the 1970s. Others felt that for the first time in their professional lives they were witnessing a true paradigm shift, a transformation in a way of thinking.

Those who recognized chaos in the early days agonized over

how to shape their thoughts and findings into publishable form. Work fell between disciplines—for example, too abstract for physicists yet too experimental for mathematicians. To some the difficulty of communicating the new ideas and the ferocious resistance from traditional quarters showed how revolutionary the new science was. Shallow ideas can be assimilated; ideas that require people to reorganize their picture of the world provoke hostility. A physicist at the Georgia Institute of Technology, Joseph Ford, started quoting Tolstoy: "I know that most men, including those at ease with problems of the greatest complexity, can seldom accept even the simplest and most obvious truth if it be such as would oblige them to admit the falsity of conclusions which they have delighted in explaining to colleagues, which they have proudly taught to others, and which they have woven, thread by thread, into the fabric of their lives."

Many mainstream scientists remained only dimly aware of the emerging science. Some, particularly traditional fluid dynamicists, actively resented it. At first, the claims made on behalf of chaos sounded wild and unscientific. And chaos relied on mathematics that seemed unconventional and difficult.

As the chaos specialists spread, some departments frowned on these somewhat deviant scholars; others advertised for more. Some journals established unwritten rules against submissions on chaos; other journals came forth to handle chaos exclusively. The chaoticists or chaologists (such coinages could be heard) turned up with disproportionate frequency on the yearly lists of important fellowships and prizes. By the middle of the eighties a process of academic diffusion had brought chaos specialists into influential positions within university bureaucracies. Centers and institutes were founded to specialize in "nonlinear dynamics" and "complex systems."

Chaos has become not just theory but also method, not just a canon of beliefs but also a way of doing science. Chaos has created its own technique of using computers, a technique that does not require the vast speed of Crays and Cybers but instead favors modest terminals that allow flexible interaction. To chaos researchers, mathematics has become an experimental science, with the computer replacing laboratories full of test tubes and microscopes. Graphic images are the key. "It's masochism for a mathematician to do without pictures," one chaos specialist would say. "How can they see the relationship between that motion and this? How can they develop intuition?" Some carry out their work explicitly denying that it is a revolution; others deliberately use Kuhn's language of paradigm shifts to describe the changes they witness.

Stylistically, early chaos papers recalled the Benjamin Franklin era in the way they went back to first principles. As Kuhn notes, established sciences take for granted a body of knowledge that serves as a communal starting point for investigation. To avoid boring their colleagues, scientists routinely begin and end their papers with esoterica. By contrast, articles on chaos from the late 1970s onward sounded evangelical, from their preambles to their perorations. They declared new credos, and they often ended with pleas for action. *These results appear to us to be both exciting and highly provocative. A theoretical picture of the transition to turbulence is just beginning to emerge. The heart of chaos is mathematically accessible. Chaos now presages the future as none will gainsay. But to accept the future, one must renounce much of the past.*

New hopes, new styles, and, most important, a new way of seeing. Revolutions do not come piecemeal. One account of nature replaces another. Old problems are seen in a new light and other problems are recognized for the first time. Something takes place that resembles a whole industry retooling for new production. In Kuhn's words, "It is rather as if the professional community had been suddenly transported to another planet where familiar objects are seen in a different light and are joined by unfamiliar ones as well."

THE LABORATORY MOUSE of the new science was the pendulum: emblem of classical mechanics, exemplar of constrained action, epitome of clockwork regularity. A bob swings free at the end of a rod. What could be further removed from the wildness of turbulence?

Where Archimedes had his bathtub and Newton his apple, so, according to the usual suspect legend, Galileo had a church lamp, swaying back and forth, time and again, on and on, sending its message monotonously into his consciousness. Christian Huygens turned the predictability of the pendulum into a means of timekeeping, sending Western civilization down a road from which there was no return. Foucault, in the Panthéon of Paris, used a twenty-story-high pendulum to demonstrate the earth's rotation. Every clock and every wristwatch (until the era of vibrating quartz) relied on a pendulum of some size or shape. (For that matter, the oscillation of quartz is not so different.) In space, free of friction, periodic motion comes from the orbits of heavenly bodies, but on earth virtually any regular oscillation comes from some cousin of the pendulum. Basic electronic circuits are described by equations exactly the same as those describing a swinging bob. The electronic oscillations are millions of times faster, but the physics is the same. By the twentieth century, though, classical mechanics was strictly a business for classrooms and routine engineering projects. Pendulums decorated science museums and enlivened airport gift shops in the form of rotating plastic "space balls." No research physicist bothered with pendulums.

Yet the pendulum still had surprises in store. It became a touchstone, as it had for Galileo's revolution. When Aristotle looked at a pendulum, he saw a weight trying to head earthward but swinging violently back and forth because it was constrained by its rope. To the modern ear this sounds foolish. For someone bound by classical concepts of motion, inertia, and gravity, it is hard to appreciate the self-consistent world view that went with Aristotle's understanding of a pendulum. Physical motion, for Aristotle, was not a quantity or a force but rather a kind of change, just as a person's growth is a kind of change. A falling weight is simply seeking its most natural state, the state it will reach if left to itself. In context, Aristotle's view made sense. When Galileo looked at a pendulum, on the other hand, he saw a regularity that could be measured. To explain it required a revolutionary way of understanding objects in motion. Galileo's advantage over the ancient Greeks was not that he had better data. On the contrary, his idea of timing a pendulum precisely was to get some friends together to count the oscillations over a twenty-four-hour period— a labor-intensive experiment. Galileo saw the regularity because he already had a theory that predicted it. He understood what Aristotle could not: that a moving object tends to keep moving, that a change in speed or direction could only be explained by some external force, like friction.

In fact, so powerful was his theory that he saw a regularity that did *not* exist. He contended that a pendulum of a given length not only keeps precise time but keeps the same time no matter how wide or narrow the angle of its swing. A wide-swinging pendulum has farther to travel, but it happens to travel just that much faster. In other words, its period remains independent of its amplitude. "If two friends shall set themselves to count the oscillations, one counting the wide ones and the other the narrow, they will see that they may count not just tens, but even hundreds, without disagreeing by even one, or part of one." Galileo phrased his claim in terms of experimentation, but the theory made it convincing—so much so that it is still taught as gospel in most high school physics courses. But it is wrong. The regularity Galileo saw is only an approximation. The changing angle of the bob's motion creates a slight nonlinearity in the equations. At low amplitudes, the error is almost nonexistent. But it is there, and it is measurable even in an experiment as crude as the one Galileo describes.

Small nonlinearities were easy to disregard. People who conduct experiments learn quickly that they live in an imperfect world. In the centuries since Galileo and Newton, the search for regularity in experiment has been fundamental. Any experimentalist looks for quantities that remain the same, or quantities that are zero. But that means disregarding bits of messiness that interfere with a neat picture. If a chemist finds two substances in a constant proportion of 2.001 one day, and 2.003 the next day, and 1.998 the day after, he would be a fool not to look for a theory that would explain a perfect two-to-one ratio.

To get his neat results, Galileo also had to disregard nonlinearities that he knew of: friction and air resistance. Air resistance is a notorious experimental nuisance, a complication that had to be stripped away to reach the essence of the new science of mechanics. Does a feather fall as rapidly as a stone? All experience with falling objects says no. The story of Galileo dropping balls off the tower of Pisa, as a piece of myth, is a story about *changing* intuitions by inventing an ideal scientific world where regularities can be separated from the disorder of experience.

To separate the effects of gravity on a given mass from the effects of air resistance was a brilliant intellectual achievement. It allowed Galileo to close in on the essence of inertia and momentum. Still, in the real world, pendulums eventually do exactly what Aristotle's quaint paradigm predicted. They stop. In laying the groundwork for the next paradigm shift, physicists began to face up to what many believed was a deficiency in their education about simple systems like the pendulum. By our century, dissipative processes like friction were recognized, and students learned to include them in equations. Students also learned that nonlinear systems were usually unsolvable, which was true, and that they tended to be exceptions—which was not true. Classical mechanics described the behavior of whole classes of moving objects, pendulums and double pendulums, coiled springs and bent rods, plucked strings and bowed strings. The mathematics applied to fluid systems and to electrical systems. But almost no one in the classical era suspected the chaos that could lurk in dynamical systems if nonlinearity was given its due.

A physicist could not truly understand turbulence or complexity unless he understood pendulums—and understood them in a way that was impossible in the first half of the twentieth century. As chaos began to unite the study of different systems, pendulum dynamics broadened to cover high technologies from lasers to superconducting Josephson junctions. Some chemical reactions displayed pendulum-like behavior, as did the beating heart. The unexpected possibilities extended, one physicist wrote, to "physiological and psychiatric medicine, economic forecasting, and perhaps the evolution of society."

Consider a playground swing. The swing accelerates on its way down, decelerates on its way up, all the while losing a bit of speed to friction. It gets a regular push—say, from some clockwork machine. All our intuition tells us that, no matter where the swing might start, the motion will eventually settle down to a regular back and forth pattern, with the swing coming to the same height each time. That can happen. Yet, odd as it seems, the motion can also turn erratic, first high, then low, never settling down to a steady state and never exactly repeating a pattern of swings that came before.

The surprising, erratic behavior comes from a nonlinear twist in the flow of energy in and out of this simple oscillator. The swing is damped and it is driven: damped because friction is trying to bring it to a halt, driven because it is getting a periodic push. Even when a damped, driven system is at equilibrium, it is not at equilibrium, and the world is full of such systems, beginning with the weather, damped by the friction of moving air and water and by the dissipation of heat to outer space, and driven by the constant push of the sun's energy.

But unpredictability was not the reason physicists and mathematicians began taking pendulums seriously again in the sixties and seventies. Unpredictability was only the attention-grabber. Those studying chaotic dynamics discovered that the disorderly behavior of simple systems acted as a *creative* process. It generated complexity: richly organized patterns, sometimes stable and sometimes unstable, sometimes finite and sometimes infinite, but always with the fascination of living things. That was why scientists played with toys.

One toy, sold under the name "Space Balls" or "Space Trapeze," is a pair of balls at opposite ends of a rod, sitting like the crossbar of a T atop a pendulum with a third, heavier ball at its foot. The lower ball swings back and forth while the upper rod rotates freely. All three balls have little magnets inside, and once set in motion the device keeps going because it has a battery-powered electromagnet embedded in the base. The device senses the approach of the lowest ball and gives it a small magnetic kick each time it passes. Sometimes the apparatus settles into a steady, rhythmic swinging. But other times, its motion seems to remain chaotic, always changing and endlessly surprising.

Another common pendulum toy is no more than a so-called spherical pendulum—a pendulum free to swing not just back and forth but in any direction. A few small magnets are placed around its base. The magnets attract the metal bob, and when the pendulum stops, it will have been captured by one of them. The idea is to set the pendulum swinging and guess which magnet will win. Even with just three magnets placed in a triangle, the pendulum's motion cannot be predicted. It will swing back and forth between A and B for a while, then switch to B and C, and then, just as it seems to be settling on C, jump back to A. Suppose a scientist systematically explores the behavior of this toy by making a map, as follows: Pick a starting point; hold the bob there and let go; color the point red, blue, or green, depending on which magnet ends up with the bob. What will the map look like? It will have regions of solid red, blue, or green, as one might expect— regions where the bob will swing reliably to a particular magnet. But it can also have regions where the colors are woven together with infinite complexity. Adjacent to a red point, no matter how close one chooses to look, no matter how much one magnifies the map, there will be green points and blue points. For all practical purposes, then, the bob's destiny will be impossible to guess.

Traditionally, a dynamicist would believe that to write down a system's equations is to understand the system. How better to capture the essential features? For a playground swing or a toy, the equations tie together the pendulum's angle, its velocity, its friction, and the force driving it. But because of the little bits of nonlinearity in these equations, a dynamicist would find himself helpless to answer the easiest practical questions about the future of the system. A computer can address the problem by simulating it, rapidly calculating each cycle. But simulation brings its own problem: the tiny imprecision built into each calculation rapidly takes over, because this is a system with sensitive dependence on initial conditions. Before long, the signal disappears and all that remains is noise.

Or is it? Lorenz had found unpredictability, but he had also found pattern. Others, too, discovered suggestions of structure amid seemingly random behavior. The example of the pendulum was simple enough to disregard, but those who chose not to dis- regard it found a provocative message. In some sense, they real- ized, physics understood perfectly the fundamental mechanisms of pendulum motion but could not extend that understanding to the long term. The microscopic pieces were perfectly clear; the macroscopic behavior remained a mystery. The tradition of look- ing at systems locally—isolating the mechanisms and then adding them together—was beginning to break down. For pendulums, for fluids, for electronic circuits, for lasers, knowledge of the funda- mental equations no longer seemed to be the right kind of knowl- edge at all.

As the 1960s went on, individual scientists made discoveries that paralleled Lorenz's: a French astronomer studying galactic orbits, for example, and a Japanese electrical engineer modeling electronic circuits. But the first deliberate, coordinated attempt to understand how global behavior might differ from local behavior came from mathematicians. Among them was Stephen Smale of the University of California at Berkeley, already famous for unraveling the most esoteric problems of many-dimensional topology. A young physicist, making small talk, asked what Smale was working on. The answer stunned him: "Oscillators." It was absurd. Oscillators—pendulums, springs, or electrical circuits—were the sort of problem that a physicist finished off early in his training. They were easy. Why would a great mathematician be studying elementary physics? Not until years later did the young man realize that Smale was looking at nonlinear oscillators, chaotic oscillators, and seeing things that physicists had learned not to see.

SMALE MADE A BAD CONJECTURE. In the most rigorous mathematical terms, he proposed that practically all dynamical systems tended to settle, most of the time, into behavior that was not too strange. As he soon learned, things were not so simple.

Smale was a mathematician who did not just solve problems but also built programs of problems for others to solve. He parlayed his understanding of history and his intuition about nature into an ability to announce, quietly, that a whole untried area of research was now worth a mathematician's time. Like a successful businessman, he evaluated risks and coolly planned his strategy, and he had a Pied Piper quality. Where Smale led, many followed. His reputation was not confined to mathematics, though. Early in the Vietnam war, he and Jerry Rubin organized "International Days of Protest" and sponsored efforts to stop the trains carrying troops through California. In 1966, while the House Un-American Activities Committee was trying to subpoena him, he was heading for Moscow to attend the International Congress of Mathematicians. There he received the Fields Medal, the highest honor of his profession.

The scene in Moscow that summer became an indelible part of the Smale legend. Five thousand agitated and agitating mathematicians had gathered. Political tensions were high. Petitions were circulating. As the conference drew toward its close, Smale responded to a request from a North Vietnamese reporter by giving a press conference on the broad steps of Moscow University. He began by condemning the American intervention in Vietnam, and then, just as his hosts began to smile, added a condemnation of the Soviet invasion of Hungary and the absence of political freedom in the Soviet Union. When he was done, he was quickly hustled away in a car for questioning by Soviet officials. When he returned to California, the National Science Foundation canceled his grant.

Smale's Fields Medal honored a famous piece of work in topology, a branch of mathematics that flourished in the twentieth century and had a particular heyday in the fifties. Topology studies the properties that remain unchanged when shapes are deformed by twisting or stretching or squeezing. Whether a shape is square or round, large or small, is irrelevant in topology, because stretching can change those properties. Topologists ask whether a shape is connected, whether it has holes, whether it is knotted. They imagine surfaces not just in the one-, two-, and three-dimensional universes of Euclid, but in spaces of many dimensions, impossible to visualize. Topology is geometry on rubber sheets. It concerns the qualitative rather than the quantitative. It asks, if you don't know the measurements, what can you say about overall structure. Smale had solved one of the historic, outstanding problems of topology, the Poincaré conjecture, for spaces of five dimensions and higher, and in so doing established a secure standing as one of the great men of the field. In the 1960s, though, he left topology for untried territory. He began studying dynamical systems.

Both subjects, topology and dynamical systems, went back to Henri Poincaré, who saw them as two sides of one coin. Poincaré, at the turn of the century, had been the last great mathematician to bring a geometric imagination to bear on the laws of motion in the physical world. He was the first to understand the possibility of chaos; his writings hinted at a sort of unpredictability almost as severe as the sort Lorenz discovered. But after Poincaré's death, while topology flourished, dynamical systems atrophied. Even the name fell into disuse; the subject to which Smale nominally turned was differential equations. Differential equations describe the way systems change continuously over time. The tradition was to look at such things locally, meaning that engineers or physicists would consider one set of possibilities at a time. Like Poincaré, Smale wanted to understand them globally, meaning that he wanted to understand the entire realm of possibilities at once.

Any set of equations describing a dynamical system—Lorenz's, for example—allows certain parameters to be set at the start. In the case of thermal convection, one parameter concerns the viscosity of the fluid. Large changes in parameters can make large differences in a system—for example, the difference between arriving at a steady state and oscillating periodically. But physicists assumed that very small changes would cause only very small differences in the numbers, not qualitative changes in behavior.

Linking topology and dynamical systems is the possibility of using a shape to help visualize the whole range of behaviors of a system. For a simple system, the shape might be some kind of curved surface; for a complicated system, a manifold of many dimensions. A single point on such a surface represents the state of a system at an instant frozen in time. As a system progresses through time, the point moves, tracing an orbit across this surface. Bending the shape a little corresponds to changing the system's parameters, making a fluid more viscous or driving a pendulum a little harder. Shapes that look roughly the same give roughly the same kinds of behavior. If you can visualize the shape, you can understand the system.

When Smale turned to dynamical systems, topology, like most pure mathematics, was carried out with an explicit disdain for real-world applications. Topology's origins had been close to physics, but for mathematicians the physical origins were forgotten and shapes were studied for their own sake. Smale fully believed in that ethos—he was the purest of the pure—yet he had an idea that the abstract, esoteric development of topology might now have something to contribute to physics, just as Poincaré had intended at the turn of the century.

One of Smale's first contributions, as it happened, was his faulty conjecture. In physical terms, he was proposing a law of nature something like this: A system can behave erratically, but the erratic behavior cannot be *stable*. Stability—"stability in the sense of Smale," as mathematicians would sometimes say—was a crucial property. Stable behavior in a system was behavior that would not disappear just because some number was changed a tiny bit. Any system could have both stable and unstable behaviors within it. The equations governing a pencil standing on its point have a good mathematical solution with the center of gravity directly above the point—but you cannot stand a pencil on its point because the solution is unstable. The slightest perturbation draws the system away from that solution. On the other hand, a marble lying at the bottom of a bowl stays there, because if the marble is perturbed slightly it rolls back. Physicists assumed that any behavior they could actually observe regularly would have to be stable, since in real systems tiny disturbances and uncertainties are unavoidable. You never know the parameters exactly. If you want a model that will be both physically realistic and robust in the face of small perturbations, physicists reasoned that you must surely want a stable model.

The bad news arrived in the mail soon after Christmas 1959, when Smale was living temporarily in an apartment in Rio de Janeiro with his wife, two infant children, and a mass of diapers. His conjecture had defined a class of differential equations, all structurally stable. Any chaotic system, he claimed, could be approximated as closely as you liked by a system in his class. It was not so. A letter from a colleague informed him that many systems were not so well-behaved as he had imagined, and it described a counterexample, a system with chaos and stability, together. This system was robust. If you perturbed it slightly, as any natural system is constantly perturbed by noise, the strangeness would not go away. Robust and strange—Smale studied the letter with a disbelief that melted away slowly.

Chaos and instability, concepts only beginning to acquire formal definitions, were not the same at all. A chaotic system could be stable if its particular brand of irregularity persisted in the face of small disturbances. Lorenz's system was an example, although years would pass before Smale heard about Lorenz. The chaos Lorenz discovered, with all its unpredictability, was as stable as a marble in a bowl. You could add noise to this system, jiggle it, stir it up, interfere with its motion, and then when everything settled down, the transients dying away like echoes in a canyon, the system would return to the same peculiar pattern of irregularity as before. It was locally unpredictable, globally stable. Real dynamical systems played by a more complicated set of rules than anyone had imagined. The example described in the letter from Smale's colleague was another simple system, discovered more than a generation earlier and all but forgotten. As it happened, it was a pendulum in disguise: an oscillating electronic circuit. It was nonlinear and it was periodically forced, just like a child on a swing.

It was just a vacuum tube, really, investigated in the twenties by a Dutch electrical engineer named Balthasar van der Pol. A modern physics student would explore the behavior of such an oscillator by looking at the line traced on the screen of an oscilloscope. Van der Pol did not have an oscilloscope, so he had to monitor his circuit by listening to changing tones in a telephone handset. He was pleased to discover regularities in the behavior as he changed the current that fed it. The tone would leap from frequency to frequency as if climbing a staircase, leaving one frequency and then locking solidly onto the next. Yet once in a while van der Pol noted something strange. The behavior sounded irregular, in a way that he could not explain. Under the circumstances he was not worried. "Often an irregular noise is heard in the telephone receivers before the frequency jumps to the next lower value," he wrote in a letter to *Nature*. "However, this is a subsidiary phenomenon." He was one of many scientists who got a glimpse of chaos but had no language to understand it. For people trying to build vacuum tubes, the frequency-locking was important. But for people trying to understand the nature of complexity, the truly interesting behavior would turn out to be the "irregular noise" created by the conflicting pulls of a higher and lower frequency.

Wrong though it was, Smale's conjecture put him directly on the track of a new way of conceiving the full complexity of dynamical systems. Several mathematicians had taken another look at the possibilities of the van der Pol oscillator, and Smale now took their work into a new realm. His only oscilloscope screen was his mind, but it was a mind shaped by his years of exploring the topological universe. Smale conceived of the entire range of possibilities in the oscillator, the entire phase space, as physicists called it. Any state of the system at a moment frozen in time was represented as a point in phase space; all the information about its position or velocity was contained in the coordinates of that point. As the system changed in some way, the point would move to a new position in phase space. As the system changed continuously, the point would trace a trajectory.

For a simple system like a pendulum, the phase space might just be a rectangle: the pendulum's angle at a given instant would determine the east-west position of a point and the pendulum's speed would determine the north-south position. For a pendulum swinging regularly back and forth, the trajectory through phase space would be a loop, around and around as the system lived through the same sequence of positions over and over again.

Smale, instead of looking at any one trajectory, concentrated on the behavior of the entire space as the system changed—as more driving energy was added, for example. His intuition leapt from the physical essence of the system to a new kind of geometrical essence. His tools were topological transformations of shapes in phase space—transformations like stretching and squeezing. Sometimes these transformations had clear physical meaning. Dissipation in a system, the loss of energy to friction, meant that the system's shape in phase space would contract like

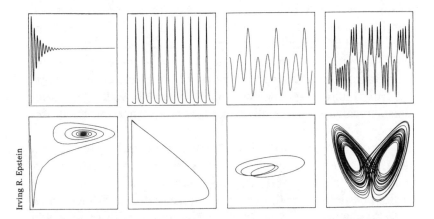

Irving R. Epstein

MAKING PORTRAITS IN PHASE SPACE. Traditional time series (*above*) and trajectories in phase space (*below*) are two ways of displaying the same data and gaining a picture of a system's long-term behavior. The first system (*left*) converges on a steady state—a point in phase space. The second repeats itself periodically, forming a cyclical orbit. The third repeats itself in a more complex waltz rhythm, a cycle with "period three." The fourth is chaotic.

a balloon losing air—finally shrinking to a point at the moment the system comes to a complete halt. To represent the full complexity of the van der Pol oscillator, he realized that the phase space would have to suffer a complex new kind of combination of transformations. He quickly turned his idea about visualizing global behavior into a new kind of model. His innovation—an enduring image of chaos in the years that followed—was a structure that became known as the horseshoe.

To make a simple version of Smale's horseshoe, you take a rectangle and squeeze it top and bottom into a horizontal bar. Take one end of the bar and fold it and stretch it around the other, making a C-shape, like a horseshoe. Then imagine the horseshoe embedded in a new rectangle and repeat the same transformation, shrinking and folding and stretching.

The process mimics the work of a mechanical taffy-maker, with rotating arms that stretch the taffy, double it up, stretch it again, and so on until the taffy's surface has become very long, very thin, and intricately self-embedded. Smale put his horseshoe through an assortment of topological paces, and, the mathematics aside, the horseshoe provided a neat visual analogue of the sen-

H. Bruce Stewart and J. M. Thompson

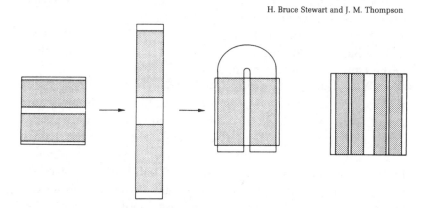

SMALE'S HORSESHOE. This topological transformation provided a basis for understanding the chaotic properties of dynamical systems. The basics are simple: A space is stretched in one direction, squeezed in another, and then folded. When the process is repeated, it produces a kind of structured mixing familiar to anyone who has rolled many-layered pastry dough. A pair of points that end up close together may have begun far apart.

sitive dependence on initial conditions that Lorenz would discover in the atmosphere a few years later. Pick two nearby points in the original space, and you cannot guess where they will end up. They will be driven arbitrarily far apart by all the folding and stretching. Afterward, two points that happen to lie nearby will have begun arbitrarily far apart.

Originally, Smale had hoped to explain all dynamical systems in terms of stretching and squeezing—with no folding, at least no folding that would drastically undermine a system's stability. But folding turned out to be necessary, and folding allowed sharp changes in dynamical behavior. Smale's horseshoe stood as the first of many new geometrical shapes that gave mathematicians and physicists a new intuition about the possibilities of motion. In some ways it was too artificial to be useful, still too much a creature of mathematical topology to appeal to physicists. But it served as a starting point. As the sixties went on, Smale assembled around him at Berkeley a group of young mathematicians who shared his excitement about this new work in dynamical systems. Another decade would pass before their work fully engaged the attention of less pure sciences, but when it did, physicists would realize that Smale had turned a whole branch of mathematics back toward the real world. It was a golden age, they said.

"It's the paradigm shift of paradigm shifts," said Ralph Abraham, a Smale colleague who became a professor of mathematics at the University of California at Santa Cruz.

"When I started my professional work in mathematics in 1960, which is not so long ago, modern mathematics in its entirety—in its entirety—was rejected by physicists, including the most avantgarde mathematical physicists. So differentiable dynamics, global analysis, manifolds of mappings, differential geometry—everything just a year or two beyond what Einstein had used—was all rejected. The romance between mathematicians and physicists had ended in divorce in the 1930s. These people were no longer speaking. They simply despised each other. Mathematical physicists refused their graduate students permission to take math courses from mathematicians: *Take mathematics from us. We will teach you what you need to know. The mathematicians are on some kind of terrible ego trip and they will destroy your mind.* That was 1960. By 1968 this had completely turned around."

Eventually physicists, astronomers, and biologists all knew they had to have the news.

A MODEST COSMIC MYSTERY: the Great Red Spot of Jupiter, a vast, swirling oval, like a giant storm that never moves and never runs down. Anyone who saw the pictures beamed across space from Voyager 2 in 1978 recognized the familiar look of turbulence on a hugely unfamiliar scale. It was one of the solar system's most venerable landmarks—"the red spot roaring like an anguished eye/ amid a turbulence of boiling eyebrows," as John Updike described it. But what *was* it? Twenty years after Lorenz, Smale, and other scientists set in motion a new way of understanding nature's flows, the other-worldly weather of Jupiter proved to be one of the many problems awaiting the altered sense of nature's possibilities that came with the science of chaos.

For three centuries it had been a case of the more you know, the less you know. Astronomers noticed a blemish on the great planet not long after Galileo first pointed his telescopes at Jupiter. Robert Hooke saw it in the 1600s. Donati Creti painted it in the Vatican's picture gallery. As a piece of coloration, the spot called for little explaining. But telescopes got better, and knowledge bred ignorance. The last century produced a steady march of theories, one on the heels of another. For example:

The Lava Flow Theory. Scientists in the late nineteenth century imagined a huge oval lake of molten lava flowing out of a volcano. Or perhaps the lava had flowed out of a hole created by a planetoid striking a thin solid crust.

The New Moon Theory. A German scientist suggested, by contrast, that the spot was a new moon on the point of emerging from the planet's surface.

The Egg Theory. An awkward new fact: the spot was seen to be drifting slightly against the planet's background. So a notion put forward in 1939 viewed the spot as a more or less solid body floating in the atmosphere the way an egg floats in water. Variations of this theory—including the notion of a drifting bubble of hydrogen or helium—remained current for decades.

The Column-of-Gas Theory. Another new fact: even though the spot drifted, somehow it never drifted far. So scientists proposed in the sixties that the spot was the top of a rising column of gas, possibly coming through a crater.

Then came Voyager. Most astronomers thought the mystery would give way as soon as they could look closely enough, and indeed, the Voyager fly-by provided a splendid album of new data, but the data, in the end, was not enough. The spacecraft pictures in 1978 revealed powerful winds and colorful eddies. In spectacular detail, astronomers saw the spot itself as a hurricane-like system of swirling flow, shoving aside the clouds, embedded in zones of east-west wind that made horizontal stripes around the planet. *Hurricane* was the best description anyone could think of, but for several reasons it was inadequate. Earthly hurricanes are powered by the heat released when moisture condenses to rain; no moist processes drive the Red Spot. Hurricanes rotate in a cyclonic direction, counterclockwise above the Equator and clockwise below, like all earthly storms; the Red Spot's rotation is anticyclonic. And most important, hurricanes die out within days.

Also, as astronomers studied the Voyager pictures, they realized that the planet was virtually all fluid in motion. They had been conditioned to look for a solid planet surrounded by a paper-thin atmosphere like earth's, but if Jupiter had a solid core anywhere, it was far from the surface. The planet suddenly looked like one big fluid dynamics experiment, and there sat the Red Spot, turning steadily around and around, thoroughly unperturbed by the chaos around it.

The spot became a gestalt test. Scientists saw what their intuitions allowed them to see. A fluid dynamicist who thought of turbulence as random and noisy had no context for understanding an island of stability in its midst. Voyager had made the mystery doubly maddening by showing small-scale features of the flow, too small to be seen by the most powerful earthbound telescopes. The small scales displayed rapid disorganization, eddies appearing and disappearing within a day or less. Yet the spot was immune. What kept it going? What kept it in place?

The National Aeronautics and Space Administration keeps its pictures in archives, a half-dozen or so around the country. One archive is at Cornell University. Nearby, in the early 1980s, Philip Marcus, a young astronomer and applied mathematician, had an office. After Voyager, Marcus was one of a half-dozen scientists in the United States and Britain who looked for ways to model the Red Spot. Freed from the ersatz hurricane theory, they found more appropriate analogues elsewhere. The Gulf Stream, for example, winding through the western Atlantic Ocean, twists and branches in subtly reminiscent ways. It develops little waves, which turn into kinks, which turn into rings and spin off from the main current—forming slow, long-lasting, anticyclonic vortices. Another parallel came from a peculiar phenomenon in meteorology known as blocking. Sometimes a system of high pressure sits offshore, slowly turning, for weeks or months, in defiance of the usual east-west flow. Blocking disrupted the global forecasting models, but it also gave the forecasters some hope, since it produced orderly features with unusual longevity.

Marcus studied those NASA pictures for hours, the gorgeous Hasselblad pictures of men on the moon and the pictures of Jupiter's turbulence. Since Newton's laws apply everywhere, Marcus programmed a computer with a system of fluid equations. To capture Jovian weather meant writing rules for a mass of dense hydrogen and helium, resembling an unlit star. The planet spins fast, each day flashing by in ten earth hours. The spin produces a strong Coriolis force, the sidelong force that shoves against a person walking across a merry-go-round, and the Coriolis force drives the spot.

Where Lorenz used his tiny model of the earth's weather to print crude lines on rolled paper, Marcus used far greater computer power to assemble striking color images. First he made contour plots. He could barely see what was going on. Then he made slides, and then he assembled the images into an animated movie. It was a revelation. In brilliant blues, reds, and yellows, a checkerboard pattern of rotating vortices coalesces into an oval with an uncanny resemblance to the Great Red Spot in NASA's animated film of the real thing. "You see this large-scale spot, happy as a clam amid the small-scale chaotic flow, and the chaotic flow is soaking up energy like a sponge," he said. "You see these little tiny filamentary structures in a background sea of chaos."

The spot is a self-organizing system, created and regulated by the same nonlinear twists that create the unpredictable turmoil around it. It is stable chaos.

As a graduate student, Marcus had learned standard physics,

solving linear equations, performing experiments designed to match linear analysis. It was a sheltered existence, but after all, nonlinear equations defy solution, so why waste a graduate student's time? Gratification was programmed into his training. As long as he kept the experiments within certain bounds, the linear approximations would suffice and he would be rewarded with the expected answer. Once in a while, inevitably, the real world would intrude, and Marcus would see what he realized years later had been the signs of chaos. He would stop and say, "Gee, what about this little fluff here." And he would be told, "Oh, it's experimental error, don't worry about it."

But unlike most physicists, Marcus eventually learned Lorenz's lesson, that a deterministic system can produce much more than just periodic behavior. He knew to look for wild disorder, and he knew that islands of structure could appear within the disorder. So he brought to the problem of the Great Red Spot an understanding that a complex system can give rise to turbulence and coherence at the same time. He could work within an emerging discipline that was creating its own tradition of using the computer as an experimental tool. And he was willing to think of himself as a new kind of scientist: not primarily an astronomer, not a fluid dynamicist, not an applied mathematician, but a specialist in chaos.

Life's Ups and Downs

The result of a mathematical development should be
continuously checked against one's own intuition
about what constitutes reasonable biological
behavior. When such a check reveals disagreement,
then the following possibilities must be considered:
a. A mistake has been made in the formal
   mathematical development;
b. The starting assumptions are incorrect and/or
   constitute a too drastic oversimplification;
c. One's own intuition about the biological field is
   inadequately developed;
d. A penetrating new principle has been discovered.

—HARVEY J. GOLD,
Mathematical Modeling
of Biological Systems

RAVENOUS FISH AND TASTY plankton. Rain forests dripping with nameless reptiles, birds gliding under canopies of leaves, insects buzzing like electrons in an accelerator. Frost belts where voles and lemmings flourish and diminish with tidy four-year periodicity in the face of nature's bloody combat. The world makes a messy laboratory for ecologists, a cauldron of five million interacting species. Or is it fifty million? Ecologists do not actually know.

Mathematically inclined biologists of the twentieth century built a discipline, ecology, that stripped away the noise and color of real life and treated populations as dynamical systems. Ecologists used the elementary tools of mathematical physics to describe life's ebbs and flows. Single species multiplying in a place where food is limited, several species competing for existence, epidemics spreading through host populations—all could be isolated, if not in laboratories then certainly in the minds of biological theorists.

In the emergence of chaos as a new science in the 1970s, ecologists were destined to play a special role. They used mathematical models, but they always knew that the models were thin approximations of the seething real world. In a perverse way, their awareness of the limitations allowed them to see the importance of some ideas that mathematicians had considered interesting oddities. If regular equations could produce irregular behavior—to an ecologist, that rang certain bells. The equations applied to population biology were elementary counterparts of the models used by physicists for their pieces of the universe. Yet the complexity of the real phenomena studied in the life sciences outstripped anything to be found in a physicist's laboratory. Biologists' mathematical models tended to be caricatures of reality, as did the models of economists, demographers, psychologists, and urban planners, when those soft sciences tried to bring rigor to their study of systems changing over time. The standards were different. To a physicist, a system of equations like Lorenz's was so simple it seemed virtually transparent. To a biologist, even Lorenz's equations seemed forbiddingly complex—three-dimensional, continuously variable, and analytically intractable.

Necessity created a different style of working for biologists. The matching of mathematical descriptions to real systems had to proceed in a different direction. A physicist, looking at a particular system (say, two pendulums coupled by a spring), begins by choosing the appropriate equations. Preferably, he looks them up in a handbook; failing that, he finds the right equations from first principles. He knows how pendulums work, and he knows about springs. Then he solves the equations, if he can. A biologist, by contrast, could never simply deduce the proper equations by just thinking about a particular animal population. He would have to gather data and try to find equations that produced similar output. What happens if you put one thousand fish in a pond with a limited food supply? What happens if you add fifty sharks that like to eat two fish per day? What happens to a virus that kills at a certain rate and spreads at a certain rate depending on population density? Scientists idealized these questions so that they could apply crisp formulas.

Often it worked. Population biology learned quite a bit about the history of life, how predators interact with their prey, how a change in a country's population density affects the spread of disease. If a certain mathematical model surged ahead, or reached equilibrium, or died out, ecologists could guess something about the circumstances in which a real population or epidemic would do the same.

One helpful simplification was to model the world in terms of discrete time intervals, like a watch hand that jerks forward second by second instead of gliding continuously. Differential equations describe processes that change smoothly over time, but differential equations are hard to compute. Simpler equations— "difference equations"—can be used for processes that jump from state to state. Fortunately, many animal populations do what they do in neat one-year intervals. Changes year to year are often more important than changes on a continuum. Unlike people, many insects, for example, stick to a single breeding season, so their generations do not overlap. To guess next spring's gypsy moth population or next winter's measles epidemic, an ecologist might only need to know the corresponding figure for this year. A year-by-year facsimile produces no more than a shadow of a system's intricacies, but in many real applications the shadow gives all the information a scientist needs.

The mathematics of ecology is to the mathematics of Steve Smale what the Ten Commandments are to the Talmud: a good set of working rules, but nothing too complicated. To describe a population changing each year, a biologist uses a formalism that a high school student can follow easily. Suppose next year's population of gypsy moths will depend entirely on this year's population. You could imagine a table listing all the specific possibilities—31,000 gypsy moths this year means 35,000 next year, and so forth. Or you could capture the relationship between all the numbers for this year and all the numbers for next year as a rule—a function. The population (x) next year is a function (F) of the population this year: $x_{next} = F(x)$. Any particular function can be drawn on a graph, instantly giving a sense of its overall shape.

In a simple model like this one, following a population through time is a matter of taking a starting figure and applying the same function again and again. To get the population for a third year, you just apply the function to the result for the second year, and so on. The whole history of the population becomes available through this process of functional iteration—a feedback loop, each year's output serving as the next year's input. Feedback can get out of hand, as it does when sound from a loudspeaker feeds back through a microphone and is rapidly amplified to an unbearable shriek. Or feedback can produce stability, as a thermostat does in regulating the temperature of a house: any temperature above a fixed point leads to cooling, and any temperature below it leads to heating.

Many different types of functions are possible. A naïve approach to population biology might suggest a function that increases the population by a certain percentage each year. That would be a linear function—$x_{next} = rx$—and it would be the classic Malthusian scheme for population growth, unlimited by food supply or moral restraint. The parameter r represents the rate of population growth. Say it is 1.1; then if this year's population is 10, next year's is 11. If the input is 20,000, the output is 22,000. The population rises higher and higher, like money left forever in a compound-interest savings account.

Ecologists realized generations ago that they would have to do better. An ecologist imagining real fish in a real pond had to find a function that matched the crude realities of life—for example, the reality of hunger, or competition. When the fish proliferate, they start to run out of food. A small fish population will grow rapidly. An overly large fish population will dwindle. Or take Japanese beetles. Every August 1 you go out to your garden and count the beetles. For simplicity's sake, you ignore birds, ignore beetle diseases, and consider only the fixed food supply. A few beetles will multiply; many will eat the whole garden and starve themselves.

In the Malthusian scenario of unrestrained growth, the linear growth function rises forever upward. For a more realistic scenario, an ecologist needs an equation with some extra term that restrains growth when the population becomes large. The most natural function to choose would rise steeply when the population is small, reduce growth to near zero at intermediate values, and crash downward when the population is very large. By repeating the process, an ecologist can watch a population settle into its long-term behavior—presumably reaching some steady state. A successful foray into mathematics for an ecologist would let him say something like this: Here's an equation; here's a variable representing reproductive rate; here's a variable representing the natural death rate; here's a variable representing the additional death rate from starvation or predation; and look—the population will rise at *this* speed until it reaches *that* level of equilibrium.

How do you find such a function? Many different equations might work, and possibly the simplest is a modification of the linear, Malthusian version: $x_{next} = rx(1 - x)$. Again, the parameter r represents a rate of growth that can be set higher or lower. The new term, $1 - x$, keeps the growth within bounds, since as x rises, $1 - x$ falls.* Anyone with a calculator could pick some starting value, pick some growth rate, and carry out the arithmetic to derive next year's population.

By the 1950s several ecologists were looking at variations of that particular equation, known as the logistic difference equation. In Australia, for example, W. E. Ricker applied it to real fisheries. Ecologists understood that the growth-rate parameter r represented an important feature of the model. In the physical systems from which these equations were borrowed, that parameter corresponded to the amount of heating, or the amount of friction, or the amount of some other messy quantity. In short, the amount of nonlinearity. In a pond, it might correspond to the fecundity of the fish, the propensity of the population not just to boom but also to bust ("biotic potential" was the dignified term). The question was, how did these different parameters affect the ultimate destiny of a changing population? The obvious answer is that a lower parameter will cause this idealized population to end up at a lower level. A higher parameter will lead to a higher steady state. This turns out to be correct for many parameters—but not all. Occasionally, researchers like Ricker surely tried parameters that were

---

*For convenience, in this highly abstract model, "population" is expressed as a fraction between zero and one, zero representing extinction, one representing the greatest conceivable population of the pond.

So begin: Choose an arbitrary value for r, say, 2.7, and a starting population of .02. One minus .02 is .98. Multiply by 0.02 and you get .0196. Multiply that by 2.7 and you get .0529. The very small starting population has more than doubled. Repeat the process, using the new population as the seed, and you get .1353. With a cheap programmable calculator, this iteration is just a matter of pushing one button over and over again. The population rises to .3159, then .5835, then .6562— the rate of increase is slowing. Then, as starvation overtakes reproduction, .6092. Then .6428, then .6199, then .6362, then .6249. The numbers seem to be bouncing back and forth, but closing in on a fixed number: .6328, .6273, .6312, .6285, .6304, .6291, .6300, .6294, .6299, .6295, .6297, .6296, .6297, .6296, .6296, .6296, .6296, .6296, .6296, .6296. Success!

In the days of pencil-and-paper arithmetic, and in the days of mechanical adding machines with hand cranks, numerical exploration never went much further.

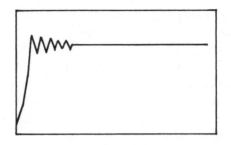

A population reaches equilibrium after rising, overshooting, and falling back.

even higher, and when they did, they must have seen chaos.

Oddly, the flow of numbers begins to misbehave, quite a nuisance for anyone calculating with a hand crank. The numbers still do not grow without limit, of course, but they do not converge to a steady level, either. Apparently, though, none of these early ecologists had the inclination or the strength to keep churning out numbers that refused to settle down. Anyway, if the population kept bouncing back and forth, ecologists assumed that it was oscillating around some underlying equilibrium. The equilibrium was the important thing. It did not occur to the ecologists that there might be *no* equilibrium.

Reference books and textbooks that dealt with the logistic equation and its more complicated cousins generally did not even acknowledge that chaotic behavior could be expected. J. Maynard Smith, in the classic 1968 *Mathematical Ideas in Biology*, gave a standard sense of the possibilities: populations often remain approximately constant or else fluctuate "with a rather regular periodicity" around a presumed equilibrium point. It wasn't that he was so naïve as to imagine that real populations could never behave erratically. He simply assumed that erratic behavior had nothing to do with the sort of mathematical models he was describing. In any case, biologists had to keep these models at arm's length. If the models started to betray their makers' knowledge of the real population's behavior, some missing feature could always explain the discrepancy: the distribution of ages in the population, some consideration of territory or geography, or the complication of having to count two sexes.

Most important, in the back of ecologists' minds was always the assumption that an erratic string of numbers probably meant that the calculator was acting up, or just lacked accuracy. The stable solutions were the interesting ones. Order was its own reward. This business of finding appropriate equations and working out the computation was hard, after all. No one wanted to waste time on a line of work that was going awry, producing no stability. And no good ecologist ever forgot that his equations were vastly oversimplified versions of the real phenomena. The whole point of oversimplifying was to model regularity. Why go to all that trouble just to see chaos?

LATER, PEOPLE WOULD SAY that James Yorke had discovered Lorenz and given the science of chaos its name. The second part was actually true.

Yorke was a mathematician who liked to think of himself as a philosopher, though this was professionally dangerous to admit. He was brilliant and soft-spoken, a mildly disheveled admirer of the mildly disheveled Steve Smale. Like everyone else, he found Smale hard to fathom. But unlike most people, he understood *why* Smale was hard to fathom. When he was just twenty-two years old, Yorke joined an interdisciplinary institute at the University of Maryland called the Institute for Physical Science and Technology, which he later headed. He was the kind of mathematician who felt compelled to put his ideas of reality to some use. He produced a report on how gonorrhea spreads that persuaded the federal government to alter its national strategies for controlling the disease. He gave official testimony to the State of Maryland during the 1970s gasoline crisis, arguing correctly (but unpersuasively) that the even-odd system of limiting gasoline sales would only make lines longer. In the era of antiwar demonstrations, when the government released a spy-plane photograph purporting to show sparse crowds around the Washington Monument at the height of a rally, he analyzed the monument's shadow to prove that the photograph had actually been taken a half-hour later, when the rally was breaking up.

At the institute, Yorke enjoyed an unusual freedom to work on problems outside traditional domains, and he enjoyed frequent contact with experts in a wide range of disciplines. One of these experts, a fluid dynamicist, had come across Lorenz's 1963 paper "Deterministic Nonperiodic Flow" in 1972 and had fallen in love with it, handing out copies to anyone who would take one. He handed one to Yorke.

Lorenz's paper was a piece of magic that Yorke had been looking for without even knowing it. It was a mathematical shock, to begin with—a chaotic system that violated Smale's original optimistic classification scheme. But it was not just mathematics; it was a vivid physical model, a picture of a fluid in motion, and Yorke knew instantly that it was a thing he wanted physicists to see. Smale had steered mathematics in the direction of such physical problems, but, as Yorke well understood, the language of mathematics remained a serious barrier to communication. If only the academic world had room for hybrid mathematician/physicists—but it did not. Even though Smale's work on dynamical systems had begun to close the gap, mathematicians continued to speak one language, physicists another. As the physicist Murray Gell-Mann once remarked: "Faculty members are familiar with a certain kind of person who looks to the mathematicians like a good physicist and looks to the physicists like a good mathematician. Very properly, they do not want that kind of person around." The standards of the two professions were different. Mathematicians proved theorems by ratiocination; physicists' proofs used heavier equipment. The objects that made up their worlds were different. Their examples were different.

Smale could be happy with an example like this: take a number, a fraction between zero and one, and double it. Then drop the integer part, the part to the left of the decimal point. Then repeat the process. Since most numbers are irrational and unpredictable in their fine detail, the process will just produce an unpredictable sequence of numbers. A physicist would see nothing there but a trite mathematical oddity, utterly meaningless, too simple and too abstract to be of use. Smale, though, knew intuitively that this mathematical trick would appear in the essence of many physical systems.

To a physicist, a legitimate example was a differential equation that could be written down in simple form. When Yorke saw Lorenz's paper, even though it was buried in a meteorology journal, he knew it was an example that physicists would understand. He gave a copy to Smale, with his address label pasted on so that Smale would return it. Smale was amazed to see that this meteorologist—*ten years earlier*—had discovered a kind of chaos that Smale himself had once considered mathematically impossible. He made many photocopies of "Deterministic Nonperiodic Flow," and thus arose the legend that Yorke had discovered Lorenz. Every copy of the paper that ever appeared in Berkeley had Yorke's address label on it.

Yorke felt that physicists had *learned* not to see chaos. In daily life, the Lorenzian quality of sensitive dependence on initial conditions lurks everywhere. A man leaves the house in the morning thirty seconds late, a flowerpot misses his head by a few millimeters, and then he is run over by a truck. Or, less dramatically, he misses a bus that runs every ten minutes—his connection to a train that runs every hour. Small perturbations in one's daily trajectory can have large consequences. A batter facing a pitched ball knows that approximately the same swing will not give approximately the same result, baseball being a game of inches. Science, though—science was different.

Pedagogically speaking, a good share of physics and mathematics was—and is—writing differential equations on a blackboard and showing students how to solve them. Differential equations represent reality as a continuum, changing smoothly from place to place and from time to time, not broken in discrete grid points or time steps. As every science student knows, solving differential equations is hard. But in two and a half centuries, scientists have built up a tremendous body of knowledge about them: handbooks and catalogues of differential equations, along with various methods for solving them, or "finding a closed-form integral," as a scientist will say. It is no exaggeration to say that the vast business of calculus made possible most of the practical triumphs of post-medieval science; nor to say that it stands as one of the most ingenious creations of humans trying to model the changeable world around them. So by the time a scientist masters this way of thinking about nature, becoming comfortable with the theory and the hard, hard practice, he is likely to have lost sight of one fact. Most differential equations cannot be solved at all.

"If you could write down the solution to a differential equation," Yorke said, "then necessarily it's not chaotic, because to write it down, you must find regular invariants, things that are conserved, like angular momentum. You find enough of these things, and that lets you write down a solution. But this is exactly the way to eliminate the possibility of chaos."

The solvable systems are the ones shown in textbooks. They behave. Confronted with a nonlinear system, scientists would have to substitute linear approximations or find some other uncertain backdoor approach. Textbooks showed students only the rare nonlinear systems that would give way to such techniques. They did not display sensitive dependence on initial conditions. Nonlinear systems with real chaos were rarely taught and rarely learned. When people stumbled across such things—and people did—all their training argued for dismissing them as aberrations. Only a few were able to remember that the solvable, orderly, linear systems were the aberrations. Only a few, that is, understood how nonlinear nature is in its soul. Enrico Fermi once exclaimed, "It does not say in the Bible that all laws of nature are expressible linearly!" The mathematician Stanislaw Ulam remarked that to call the study of chaos "nonlinear science" was like calling zoology "the study of nonelephant animals."

Yorke understood. "The first message is that there is disorder. Physicists and mathematicians want to discover regularities. People say, what use is disorder. But people have to know about disorder if they are going to deal with it. The auto mechanic who doesn't know about sludge in valves is not a good mechanic." Scientists and nonscientists alike, Yorke believed, can easily mislead themselves about complexity if they are not properly attuned to it. Why do investors insist on the existence of cycles in gold and silver prices? Because periodicity is the most complicated orderly behavior they can imagine. When they see a complicated pattern of prices, they look for some periodicity wrapped in a little random noise. And scientific experimenters, in physics or chemistry or biology, are no different. "In the past, people have seen chaotic behavior in innumerable circumstances," Yorke said. "They're running a physical experiment, and the experiment behaves in an erratic manner. They try to fix it or they give up. They explain the erratic behavior by saying there's noise, or just that the experiment is bad."

Yorke decided there was a message in the work of Lorenz and Smale that physicists were not hearing. So he wrote a paper for the most broadly distributed journal he thought he could publish in, the *American Mathematical Monthly*. (As a mathematician, he found himself helpless to phrase ideas in a form that physics journals would find acceptable; it was only years later that he hit upon the trick of collaborating with physicists.) Yorke's paper was important on its merits, but in the end its most influential feature was its mysterious and mischievous title: "Period Three Implies Chaos." His colleagues advised him to choose something more sober, but Yorke stuck with a word that came to stand for the whole growing business of deterministic disorder. He also talked to his friend Robert May, a biologist.

MAY CAME TO BIOLOGY through the back door, as it happened. He started as a theoretical physicist in his native Sydney, Australia, the son of a brilliant barrister, and he did postdoctoral work in applied mathematics at Harvard. In 1971, he went for a year to the Institute for Advanced Study in Princeton; instead of doing the work he was supposed to be doing, he found himself drifting over to Princeton University to talk to the biologists there.

Even now, biologists tend not to have much mathematics beyond calculus. People who like mathematics and have an aptitude for it tend more toward mathematics or physics than the life sciences. May was an exception. His interests at first tended toward the abstract problems of stability and complexity, mathematical explanations of what enables competitors to coexist. But he soon began to focus on the simplest ecological questions of how single populations behave over time. The inevitably simple models seemed less of a compromise. By the time he joined the Princeton faculty for good—eventually he would become the university's dean for research—he had already spent many hours studying a version of the logistic difference equation, using mathematical analysis and also a primitive hand calculator.

Once, in fact, on a corridor blackboard back in Sydney, he wrote the equation out as a problem for the graduate students. It was starting to annoy him. *"What the Christ happens when lambda gets bigger than the point of accumulation?"* What happened, that is, when a population's rate of growth, its tendency toward boom and bust, passed a critical point. By trying different values of this nonlinear parameter, May found that he could dramatically change the system's character. Raising the parameter meant raising the degree of nonlinearity, and that changed not just the quantity of the outcome, but also its quality. It affected not just the final population at equilibrium, but also whether the population would reach equilibrium at all.

When the parameter was low, May's simple model settled on a steady state. When the parameter was high, the steady state would break apart, and the population would oscillate between two alternating values. When the parameter was very high, the system—*the very same system*—seemed to behave unpredictably. Why? What exactly happened at the boundaries between the different kinds of behavior? May couldn't figure it out. (Nor could the graduate students.)

May carried out a program of intense numerical exploration into the behavior of this simplest of equations. His program was analogous to Smale's: he was trying to understand this one simple equation *all at once,* not locally but globally. The equation was far simpler than anything Smale had studied. It seemed incredible that its possibilities for creating order and disorder had not been exhausted long since. But they had not. Indeed, May's program was just a beginning. He investigated hundreds of different values of the parameter, setting the feedback loop in motion and watching to see where—and whether—the string of numbers would settle down to a fixed point. He focused more and more closely on the critical boundary between steadiness and oscillation. It was as if he had his own fish pond, where he could wield fine mastery over the "boom-and-bustiness" of the fish. Still using the logistic equation, $x_{next} = rx(1-x)$, May increased the parameter as slowly as he could. If the parameter was 2.7, then the population would be .6292. As the parameter rose, the final population rose slightly, too, making a line that rose slightly as it moved from left to right on the graph.

Suddenly, though, as the parameter passed 3, the line broke in two. May's imaginary fish population refused to settle down to a single value, but oscillated between two points in alternating years. Starting at a low number, the population would rise and

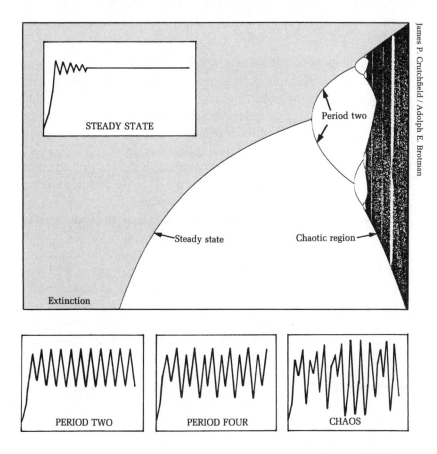

STEADY STATE

Period two

Steady state

Chaotic region

Extinction

PERIOD TWO

PERIOD FOUR

CHAOS

PERIOD-DOUBLINGS AND CHAOS. Instead of using individual diagrams to show the behavior of populations with different degrees of fertility, Robert May and other scientists used a "bifurcation diagram" to assemble all the information into a single picture.

The diagram shows how changes in one parameter—in this case, a wildlife population's "boom-and-bustiness"—would change the ultimate behavior of this simple system. Values of the parameter are represented from left to right; the final population is plotted on the vertical axis. In a sense, turning up the parameter value means driving a system harder, increasing its nonlinearity.

Where the parameter is low (left), the population becomes extinct. As the parameter rises (center), so does the equilibrium level of the population. Then, as the parameter rises further, the equilibrium splits in two, just as turning up the heat in a convecting fluid causes an instability to set in; the population begins to alternate between two different levels. The splittings, or bifurcations, come faster and faster. Then the system turns chaotic (right), and the population visits infinitely many differe͠ values. (For a blowup of the chaotic region, see pages 74–75.)

then fluctuate until it was steadily flipping back and forth. Turning up the knob a bit more—raising the parameter a bit more—would split the oscillation again, producing a string of numbers that settled down to four different values, each returning every fourth year.* Now the population rose and fell on a regular four-year schedule. The cycle had doubled again—first from yearly to every two years, and now to four. Once again, the resulting cyclical behavior was stable; different starting values for the population would converge on the same four-year cycle.

As Lorenz had discovered a decade before, the only way to make sense of such numbers and preserve one's eyesight is to create a graph. May drew a sketchy outline meant to sum up all the knowledge about the behavior of such a system at different parameters. The level of the parameter was plotted horizontally, increasing from left to right. The population was represented vertically. For each parameter, May plotted a point representing the final outcome, after the system reached equilibrium. At the left, where the parameter was low, this outcome would just be a point, so different parameters produced a line rising slightly from left to right. When the parameter passed the first critical point, May would have to plot two populations: the line would split in two, making a sideways Y or a pitchfork. This split corresponded to a population going from a one-year cycle to a two-year cycle.

As the parameter rose further, the number of points doubled again, then again, then again. It was dumbfounding—such complex behavior, and yet so tantalizingly regular. "The snake in the mathematical grass" was how May put it. The doublings themselves were bifurcations, and each bifurcation meant that the pattern of repetition was breaking down a step further. A population that had been stable would alternate between different levels every other year. A population that had been alternating on a two-year

---

*With a parameter of 3.5, say, and a starting value of .4, he would see a string of numbers like this: .4000, .8400, .4704, .8719,
 .3908, .8332, .4862, .8743,
 .3846, .8284, .4976, .8750,
 .3829, .8270, .4976, .8750,
 .3829, .8270, .5008, .8750,
 .3828, .8269, .5009, .8750,
 .3828, .8269, .5009, .8750, etc.

cycle would now vary on the third and fourth years, thus switching to period four.

These bifurcations would come faster and faster—4, 8, 16, 32 . . . —and suddenly break off. Beyond a certain point, the "point of accumulation," periodicity gives way to chaos, fluctuations that never settle down at all. Whole regions of the graph are completely blacked in. If you were following an animal population governed by this simplest of nonlinear equations, you would think the changes from year to year were absolutely random, as though blown about by environmental noise. Yet in the middle of this complexity, stable cycles suddenly return. Even though the parameter is rising, meaning that the nonlinearity is driving the system harder and harder, a window will suddenly appear with a regular period: an odd period, like 3 or 7. The pattern of changing population repeats itself on a three-year or seven-year cycle. Then the period-doubling bifurcations begin all over at a faster rate, rapidly passing through cycles of 3, 6, 12 . . . or 7, 14, 28 . . . , and then breaking off once again to renewed chaos.

At first, May could not see this whole picture. But the fragments he could calculate were unsettling enough. In a real-world system, an observer would see just the vertical slice corresponding to one parameter at a time. He would see only one kind of behavior—possibly a steady state, possibly a seven-year cycle, possibly apparent randomness. He would have no way of knowing that the same system, with some slight change in some parameter, could display patterns of a completely different kind.

James Yorke analyzed this behavior with mathematical rigor in his "Period Three Implies Chaos" paper. He proved that in *any* one-dimensional system, if a regular cycle of period three ever appears, then the same system will also display regular cycles of every other length, as well as completely chaotic cycles. This was the discovery that came as an "electric shock" to physicists like Freeman Dyson. It was so contrary to intuition. You would think it would be trivial to set up a system that would repeat itself in a period-three oscillation without ever producing chaos. Yorke showed that it was impossible.

Startling though it was, Yorke believed that the public relations value of his paper outweighed the mathematical substance. That was partly true. A few years later, attending an international

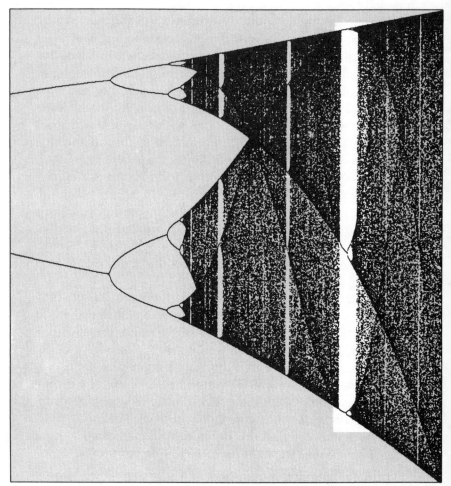

WINDOWS OF ORDER INSIDE CHAOS. Even with the simplest equation, the region of chaos in a bifurcation diagram proves to have an intricate structure—far more orderly than Robert May could guess at first. First, the bifurcations produce periods of 2, 4, 8, 16. . . . Then chaos begins, with no regular periods. But then, as the system is driven harder, windows appear with odd periods. A stable period 3 appears (blowup, *top right*), and then the period-doubling begins again 6, 12, 24. . . . The structure is infinitely deep. When portions are magnified (such as the middle piece of the period 3 window, *bottom right*), they turn out to resemble the whole diagram.

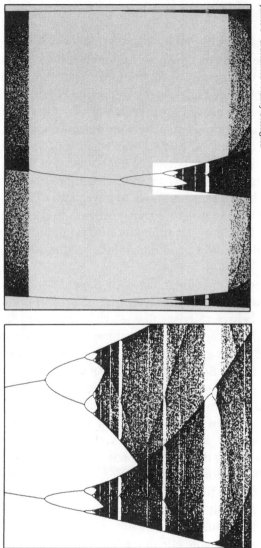

conference in East Berlin, he took some time out for sightseeing and went for a boat ride on the Spree. Suddenly he was approached by a Russian trying urgently to communicate something. With the help of a Polish friend, Yorke finally understood that the Russian was claiming to have proved the same result. The Russian refused to give details, saying only that he would send his paper. Four months later it arrived. A. N. Sarkovskii had indeed been there first, in a paper titled "Coexistence of Cycles of a Continuous Map of a Line into Itself." But Yorke had offered more than a mathematical result. He had sent a message to physicists: Chaos is ubiquitous; it is stable; it is structured. He also gave reason to believe that complicated systems, traditionally modeled by hard continuous differential equations, could be understood in terms of easy discrete maps.

The sightseeing encounter between these frustrated, gesticulating mathematicians was a symptom of a continuing communications gap between Soviet and Western science. Partly because of language, partly because of restricted travel on the Soviet side, sophisticated Western scientists have often repeated work that already existed in the Soviet literature. The blossoming of chaos in the United States and Europe has inspired a huge body of parallel work in the Soviet Union; on the other hand, it also inspired considerable bewilderment, because much of the new science was not so new in Moscow. Soviet mathematicians and physicists had a strong tradition in chaos research, dating back to the work of A. N. Kolmogorov in the fifties. Furthermore, they had a tradition of working together that had survived the divergence of mathematics and physics elsewhere.

Thus Soviet scientists were receptive to Smale—his horseshoe created a considerable stir in the sixties. A brilliant mathematical physicist, Yasha Sinai, quickly translated similar systems into thermodynamic terms. Similarly, when Lorenz's work finally reached Western physics in the seventies, it simultaneously spread in the Soviet Union. And in 1975, as Yorke and May struggled to capture the attention of their colleagues, Sinai and others rapidly assembled a powerful working group of physicists centered in Gorki. In recent years, some Western chaos experts have made a point of traveling regularly to the Soviet Union to stay current;

most, however, have had to content themselves with the Western version of their science.

In the West, Yorke and May were the first to feel the full shock of period-doubling and to pass the shock along to the community of scientists. The few mathematicians who had noted the phenomenon treated it as a technical matter, a numerical oddity: almost a kind of game playing. Not that they considered it trivial. But they considered it a thing of their special universe.

Biologists had overlooked bifurcations on the way to chaos because they lacked mathematical sophistication and because they lacked the motivation to explore disorderly behavior. Mathematicians had seen bifurcations but had moved on. May, a man with one foot in each world, understood that he was entering a domain that was astonishing and profound.

To SEE DEEPER INTO this simplest of systems, scientists needed greater computing power. Frank Hoppensteadt, at New York University's Courant Institute of Mathematical Sciences, had so powerful a computer that he decided to make a movie.

Hoppensteadt, a mathematician who later developed a strong interest in biological problems, fed the logistic nonlinear equation through his Control Data 6600 hundreds of millions of times. He took pictures from the computer's display screen at each of a thousand different values of the parameter, a thousand different tunings. The bifurcations appeared, then chaos—and then, within the chaos, the little spikes of order, ephemeral in their instability. Fleeting bits of periodic behavior. Staring at his own film, Hoppensteadt felt as if he were flying through an alien landscape. One instant it wouldn't look chaotic at all. The next instant it would be filled with unpredictable tumult. The feeling of astonishment was something Hoppensteadt never got over.

May saw Hoppensteadt's movie. He also began collecting analogues from other fields, such as genetics, economics, and fluid dynamics. As a town crier for chaos, he had two advantages over the pure mathematicians. One was that, for him, the simple equations could *not* represent reality perfectly. He knew they were just metaphors—so he began to wonder how widely the metaphors

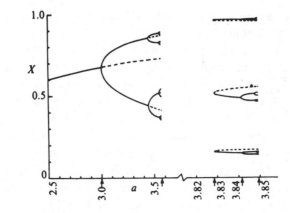

The outline of the bifurcation diagram as May first saw it, before more powerful computation revealed its rich structure.

could apply. The other was that the revelations of chaos fed directly into a vehement controversy in his chosen field.

Population biology had long been a magnet for controversy anyway. There was tension in biology departments, for example, between molecular biologists and ecologists. The molecular biologists thought that they did *real* science, crisp, hard problems, whereas the work of ecologists was vague. Ecologists believed that the technical masterpieces of molecular biology were just clever elaborations of well-defined problems.

Within ecology itself, as May saw it, a central controversy in the early 1970s dealt with the nature of population change. Ecologists were divided almost along lines of personality. Some read the message of the world to be orderly: populations are regulated and steady—with exceptions. Others read the opposite message: populations fluctuate erratically—with exceptions. By no coincidence, these opposing camps also divided over the application of hard mathematics to messy biological questions. Those who believed that populations were steady argued that they must be regulated by some deterministic mechanisms. Those who believed that populations were erratic argued that they must be bounced around by unpredictable environmental factors, wiping out whatever deterministic signal might exist. Either deterministic mathematics produced steady behavior, or random external noise produced random behavior. That was the choice.

In the context of that debate, chaos brought an astonishing message: simple deterministic models could produce what looked like random behavior. The behavior actually had an exquisite fine structure, yet any piece of it seemed indistinguishable from noise. The discovery cut through the heart of the controversy.

As May looked at more and more biological systems through the prism of simple chaotic models, he continued to see results that violated the standard intuition of practitioners. In epidemiology, for example, it was well known that epidemics tend to come in cycles, regular or irregular. Measles, polio, rubella—all rise and fall in frequency. May realized that the oscillations could be reproduced by a nonlinear model and he wondered what would happen if such a system received a sudden kick—a perturbation of the kind that might correspond to a program of inoculation. Naïve intuition suggests that the system will change smoothly in the desired direction. But actually, May found, huge oscillations are likely to begin. Even if the long-term trend was turned solidly downward, the path to a new equilibrium would be interrupted by surprising peaks. In fact, in data from real programs, such as a campaign to wipe out rubella in Britain, doctors had seen oscillations just like those predicted by May's model. Yet any health official, seeing a sharp short-term rise in rubella or gonorrhea, would assume that the inoculation program had failed.

Within a few years, the study of chaos gave a strong impetus to theoretical biology, bringing biologists and physicists into scholarly partnerships that were inconceivable a few years before. Ecologists and epidemiologists dug out old data that earlier scientists had discarded as too unwieldy to handle. Deterministic chaos was found in records of New York City measles epidemics and in two hundred years of fluctuations of the Canadian lynx population, as recorded by the trappers of the Hudson's Bay Company. Molecular biologists began to see proteins as systems in motion. Physiologists looked at organs not as static structures but as complexes of oscillations, some regular and some irregular.

All through science, May knew, specialists had seen and argued about the complex behavior of systems. Each discipline considered its particular brand of chaos to be special unto itself. The thought inspired despair. Yet what if apparent randomness could come from simple models? And what if the *same* simple models applied to complexity in different fields? May realized that the astonishing structures he had barely begun to explore had no intrinsic connection to biology. He wondered how many other sorts of scientists would be as astonished as he. He set to work on what he eventually thought of as his "messianic" paper, a review article in 1976 for *Nature*.

The world would be a better place, May argued, if every young student were given a pocket calculator and encouraged to play with the logistic difference equation. That simple calculation, which he laid out in fine detail in the *Nature* article, could counter the distorted sense of the world's possibilities that comes from a standard scientific education. It would change the way people thought about everything from the theory of business cycles to the propagation of rumors.

Chaos should be taught, he argued. It was time to recognize that the standard education of a scientist gave the wrong impression. No matter how elaborate linear mathematics could get, with its Fourier transforms, its orthogonal functions, its regression techniques, May argued that it inevitably misled scientists about their overwhelmingly nonlinear world. "The mathematical intuition so developed ill equips the student to confront the bizarre behaviour exhibited by the simplest of discrete nonlinear systems," he wrote.

"Not only in research, but also in the everyday world of politics and economics, we would all be better off if more people realized that simple nonlinear systems do not necessarily possess simple dynamical properties."

A Geometry
of Nature

And yet relation appears,
A small relation expanding like the shade
Of a cloud on sand, a shape on the side of a hill.
                    —WALLACE STEVENS,
                    "Connoisseur of Chaos"

A PICTURE OF REALITY built up over the years in Benoit Mandelbrot's mind. In 1960, it was a ghost of an idea, a faint, unfocused image. But Mandelbrot recognized it when he saw it, and there it was on the blackboard in Hendrik Houthakker's office.

Mandelbrot was a mathematical jack-of-all-trades who had been adopted and sheltered by the pure research wing of the International Business Machines Corporation. He had been dabbling in economics, studying the distribution of large and small incomes in an economy. Houthakker, a Harvard economics professor, had invited Mandelbrot to give a talk, and when the young mathematician arrived at Littauer Center, the stately economics building just north of Harvard Yard, he was startled to see his findings already charted on the older man's blackboard. Mandelbrot made a querulous joke—*how should my diagram have materialized ahead of my lecture?*—but Houthakker didn't know what Mandelbrot was talking about. The diagram had nothing to do with income distribution; it represented eight years of cotton prices.

From Houthakker's point of view, too, there was something strange about this chart. Economists generally assumed that the price of a commodity like cotton danced to two different beats, one orderly and one random. Over the long term, prices would be driven steadily by real forces in the economy—the rise and fall of the New England textile industry, or the opening of international trade routes. Over the short term, prices would bounce around more or less randomly. Unfortunately, Houthakker's data failed to match his expectations. There were too many large jumps. Most price changes were small, of course, but the ratio of small changes to large was not as high as he had expected. The distribution did not fall off quickly enough. It had a long tail.

The standard model for plotting variation was and is the bell-shaped curve. In the middle, where the hump of the bell rises, most data cluster around the average. On the sides, the low and high extremes fall off rapidly. A statistician uses a bell-shaped curve the way an internist uses a stethoscope, as the instrument of first resort. It represents the standard, so-called Gaussian distribution of things—or, simply, the normal distribution. It makes a statement about the nature of randomness. The point is that when things vary, they try to stay near an average point and they manage to scatter around the average in a reasonably smooth way. But as a means of finding paths through the economic wilderness, the standard notions left something to be desired. As the Nobel laureate Wassily Leontief put it, "In no field of empirical inquiry has so massive and sophisticated a statistical machinery been used with such indifferent results."

No matter how he plotted them, Houthakker could not make the changes in cotton prices fit the bell-shaped model. But they made a picture whose silhouette Mandelbrot was beginning to see in surprisingly disparate places. Unlike most mathematicians, he confronted problems by depending on his intuition about patterns and shapes. He mistrusted analysis, but he trusted his mental pictures. And he already had the idea that other laws, with different behavior, could govern random, stochastic phenomena. When he went back to the giant IBM research center in Yorktown Heights,

W. J. Youden

THE
NORMAL
LAW OF ERROR
STANDS OUT IN THE
EXPERIENCE OF MANKIND
AS ONE OF THE BROADEST
GENERALIZATIONS OF NATURAL
PHILOSOPHY ◆ IT SERVES AS THE
GUIDING INSTRUMENT IN RESEARCHES
IN THE PHYSICAL AND SOCIAL SCIENCES AND
IN MEDICINE AGRICULTURE AND ENGINEERING ◆
IT IS AN INDISPENSABLE TOOL FOR THE ANALYSIS AND THE
INTERPRETATION OF THE BASIC DATA OBTAINED BY OBSERVATION AND EXPERIMENT

THE BELL-SHAPED CURVE.

New York, in the hills of northern Westchester County, he carried Houthakker's cotton data in a box of computer cards. Then he sent to the Department of Agriculture in Washington for more, dating back to 1900.

Like scientists in other fields, economists were crossing the threshold into the computer era, slowly realizing that they would have the power to collect and organize and manipulate information on a scale that had been unimaginable before. Not all kinds of information were available, though, and information that could be rounded up still had to be turned into some usable form. The keypunch era was just beginning, too. In the hard sciences, investigators found it easier to amass their thousands or millions of data points. Economists, like biologists, dealt with a world of willful living beings. Economists studied the most elusive creatures of all.

But at least the economists' environment produced a constant supply of numbers. From Mandelbrot's point of view, cotton prices made an ideal data source. The records were complete and they were old, dating back continuously a century or more. Cotton was a piece of the buying-and-selling universe with a centralized market—and therefore centralized record-keeping—because at the turn of the century all the South's cotton flowed through the New York exchange on route to New England, and Liverpool's prices were linked to New York's as well.

Although economists had little to go on when it came to analyzing commodity prices or stock prices, that did not mean they lacked a fundamental viewpoint about how price changes worked. On the contrary, they shared certain articles of faith. One was a conviction that small, transient changes had nothing in common with large, long-term changes. Fast fluctuations come randomly. The small-scale ups and downs during a day's transactions are just noise, unpredictable and uninteresting. Long-term changes, however, are a different species entirely. The broad swings of prices over months or years or decades are determined by deep macroeconomic forces, the trends of war or recession, forces that should in theory give way to understanding. On the one hand, the buzz of short-term fluctuation; on the other, the signal of long-term change.

As it happened, that dichotomy had no place in the picture

of reality that Mandelbrot was developing. Instead of separating tiny changes from grand ones, his picture bound them together. He was looking for patterns not at one scale or another, but across every scale. It was far from obvious how to draw the picture he had in mind, but he knew there would have to be a kind of symmetry, not a symmetry of right and left or top and bottom but rather a symmetry of large scales and small.

Indeed, when Mandelbrot sifted the cotton-price data through IBM's computers, he found the astonishing results he was seeking. The numbers that produced aberrations from the point of view of normal distribution produced symmetry from the point of view of scaling. Each particular price change was random and unpredictable. But the sequence of changes was independent of scale: curves for daily price changes and monthly price changes matched perfectly. Incredibly, analyzed Mandelbrot's way, the degree of variation had remained constant over a tumultuous sixty-year period that saw two World Wars and a depression.

Within the most disorderly reams of data lived an unexpected kind of order. Given the arbitrariness of the numbers he was examining, why, Mandelbrot asked himself, should any law hold at all? And why should it apply equally well to personal incomes and cotton prices?

In truth, Mandelbrot's background in economics was as meager as his ability to communicate with economists. When he published an article on his findings, it was preceded by an explanatory article by one of his students, who repeated Mandelbrot's material in economists' English. Mandelbrot moved on to other interests. But he took with him a growing determination to explore the phenomenon of scaling. It seemed to be a quality with a life of its own—a signature.

INTRODUCED FOR A LECTURE years later (". . . taught economics at Harvard, engineering at Yale, physiology at the Einstein School of Medicine . . ."), he remarked proudly: "Very often when I listen to the list of my previous jobs I wonder if I exist. The intersection of such sets is surely empty." Indeed, since his early days at IBM, Mandelbrot has failed to exist in a long list of different fields. He was always an outsider, taking an unorthodox approach to an unfashionable corner of mathematics, exploring disciplines in which he was rarely welcomed, hiding his grandest ideas in efforts to get his papers published, surviving mainly on the confidence of his employers in Yorktown Heights. He made forays into fields like economics and then withdrew, leaving behind tantalizing ideas but rarely well-founded bodies of work.

In the history of chaos, Mandelbrot made his own way. Yet the picture of reality that was forming in his mind in 1960 evolved from an oddity into a full-fledged geometry. To the physicists expanding on the work of people like Lorenz, Smale, Yorke, and May, this prickly mathematician remained a sideshow—but his techniques and his language became an inseparable part of their new science.

The description would not have seemed apt to anyone who knew him in his later years, with his high imposing brow and his list of titles and honors, but Benoit Mandelbrot is best understood as a refugee. He was born in Warsaw in 1924 to a Lithuanian Jewish family, his father a clothing wholesaler, his mother a dentist. Alert to geopolitical reality, the family moved to Paris in 1936, drawn in part by the presence of Mandelbrot's uncle, Szolem Mandelbrojt, a mathematician. When the war came, the family stayed just ahead of the Nazis once again, abandoning everything but a few suitcases and joining the stream of refugees who clogged the roads south from Paris. They finally reached the town of Tulle.

For a while Benoit went around as an apprentice toolmaker, dangerously conspicuous by his height and his educated background. It was a time of unforgettable sights and fears, yet later he recalled little personal hardship, remembering instead the times he was befriended in Tulle and elsewhere by schoolteachers, some of them distinguished scholars, themselves stranded by the war. In all, his schooling was irregular and discontinuous. He claimed never to have learned the alphabet or, more significantly, multiplication tables past the fives. Still, he had a gift.

When Paris was liberated, he took and passed the month-long oral and written admissions examination for École Normale and École Polytechnique, despite his lack of preparation. Among other elements, the test had a vestigial examination in drawing, and Mandelbrot discovered a latent facility for copying the Venus de Milo. On the mathematical sections of the test—exercises in formal algebra and integrated analysis—he managed to hide his lack of training with the help of his geometrical intuition. He had realized that, given an analytic problem, he could almost always think of it in terms of some shape in his mind. Given a shape, he could find ways of transforming it, altering its symmetries, making it more harmonious. Often his transformations led directly to a solution of the analogous problem. In physics and chemistry, where he could not apply geometry, he got poor grades. But in mathematics, questions he could never have answered using proper techniques melted away in the face of his manipulations of shapes.

The École Normale and École Polytechnique were elite schools with no parallel in American education. Together they prepared fewer than 300 students in each class for careers in the French universities and civil service. Mandelbrot began in Normale, the smaller and more prestigious of the two, but left within days for Polytechnique. He was already a refugee from Bourbaki.

Perhaps nowhere but in France, with its love of authoritarian academies and received rules for learning, could Bourbaki have arisen. It began as a club, founded in the unsettled wake of World War I by Szolem Mandelbrot and a handful of other insouciant young mathematicians looking for a way to rebuild French mathematics. The vicious demographics of war had left an age gap between university professors and students, disrupting the tradition of academic continuity, and these brilliant young men set out to establish new foundations for the practice of mathematics. The name of their group was itself an inside joke, borrowed for its strange and attractive sound—so it was later guessed—from a nineteenth-century French general of Greek origin. Bourbaki was born with a playfulness that soon disappeared.

Its members met in secrecy. Indeed, not all their names are known. Their number was fixed. When one member left, as was required at age 50, another would be elected by the remaining group. They were the best and the brightest of mathematicians, and their influence soon spread across the continent.

In part, Bourbaki began in reaction to Poincaré, the great man of the late nineteenth century, a phenomenally prolific thinker and writer who cared less than some for rigor. Poincaré would say, I know it must be right, so why should I prove it? Bourbaki believed that Poincaré had left a shaky basis for mathematics, and the group began to write an enormous treatise, more and more fanatical in style, meant to set the discipline straight. Logical analysis was central. A mathematician had to begin with solid first principles and deduce all the rest from them. The group stressed the primacy of mathematics among sciences, and also insisted upon a detachment from other sciences. Mathematics was mathematics—it could not be valued in terms of its application to real physical phenomena. And above all, Bourbaki rejected the use of pictures. A mathematician could always be fooled by his visual apparatus. Geometry was untrustworthy. Mathematics should be pure, formal, and austere.

Nor was this strictly a French development. In the United States, too, mathematicians were pulling away from the demands of the physical sciences as firmly as artists and writers were pulling away from the demands of popular taste. A hermetic sensibility prevailed. Mathematicians' subjects became self-contained; their method became formally axiomatic. A mathematician could take pride in saying that his work explained nothing in the world or in science. Much good came of this attitude, and mathematicians treasured it. Steve Smale, even while he was working to reunite mathematics and natural science, believed, as deeply as he believed anything, that *mathematics should be something all by itself*. With self-containment came clarity. And clarity, too, went hand in hand with the rigor of the axiomatic method. Every serious mathematician understands that rigor is the defining strength of the discipline, the steel skeleton without which all would collapse. Rigor is what allows mathematicians to pick up a line of thought that extends over centuries and continue it, with a firm guarantee.

Even so, the demands of rigor had unintended consequences for mathematics in the twentieth century. The field develops through a special kind of evolution. A researcher picks up a problem and begins by making a decision about which way to continue. It happened that often that decision involved a choice between a path that was mathematically feasible and a path that was interesting from the point of view of understanding nature. For a mathematician, the choice was clear: he would abandon any obvious connection with nature for a while. Eventually his students would face a similar choice and make a similar decision.

Nowhere were these values as severely codified as in France, and there Bourbaki succeeded as its founders could not have imagined. Its precepts, style, and notation became mandatory. It achieved the unassailable rightness that comes from controlling all the best students and producing a steady flow of successful mathematics. Its dominance over École Normale was total and, to Mandelbrot, unbearable. He fled Normale because of Bourbaki, and a decade later he fled France for the same reason, taking up residence in the United States. Within a few decades, the relentless abstraction of Bourbaki would begin to die of a shock brought on by the computer, with its power to feed a new mathematics of the eye. But that was too late for Mandelbrot, unable to live by Bourbaki's formalisms and unwilling to abandon his geometrical intuition.

ALWAYS A BELIEVER in creating his own mythology, Mandelbrot appended this statement to his entry in Who's Who: "Science would be ruined if (like sports) it were to put competition above everything else, and if it were to clarify the rules of competition by withdrawing entirely into narrowly defined specialties. The rare scholars who are nomads-by-choice are essential to the intellectual welfare of the settled disciplines." This nomad-by-choice, who also called himself a pioneer-by-necessity, withdrew from academe when he withdrew from France, accepting the shelter of IBM's Thomas J. Watson Research Center. In a thirty-year journey from obscurity to eminence, he never saw his work embraced by the many disciplines toward which he directed it. Even mathematicians would say, without apparent malice, that whatever Mandelbrot was, he was not one of them.

He found his way slowly, always abetted by an extravagant knowledge of the forgotten byways of scientific history. He ventured into mathematical linguistics, explaining a law of the distribution of words. (Apologizing for the symbolism, he insisted that the problem came to his attention from a book review that he retrieved from a pure mathematician's wastebasket so he would have something to read on the Paris subway.) He investigated game theory. He worked his way in and out of economics. He wrote about scaling regularities in the distribution of large and small cities. The general framework that tied his work together remained in the background, incompletely formed.

Early in his time at IBM, soon after his study of commodity prices, he came upon a practical problem of intense concern to his corporate patron. Engineers were perplexed by the problem of noise in telephone lines used to transmit information from computer to computer. Electric current carries the information in discrete packets, and engineers knew that the stronger they made the current the better it would be at drowning out noise. But they found that some spontaneous noise could never be eliminated. Once in a while it would wipe out a piece of signal, creating an error.

Although by its nature the transmission noise was random, it was well known to come in clusters. Periods of errorless communication would be followed by periods of errors. By talking to the engineers, Mandelbrot soon learned that there was a piece of folklore about the errors that had never been written down, because it matched none of the standard ways of thinking: the more closely they looked at the clusters, the more complicated the patterns of errors seemed. Mandelbrot provided a way of describing the distribution of errors that predicted exactly the observed patterns. Yet it was exceedingly peculiar. For one thing, it made it impossible to calculate an *average* rate of errors—an average number of errors per hour, or per minute, or per second. On *average*, in Mandelbrot's scheme, errors approached infinite sparseness.

His description worked by making deeper and deeper separations between periods of clean transmission and periods of errors. Suppose you divided a day into hours. An hour might pass with no errors at all. Then an hour might contain errors. Then an hour might pass with no errors.

But suppose you then divided the hour with errors into smaller periods of twenty minutes. You would find that here, too, some periods would be completely clean, while some would contain a burst of errors. In fact, Mandelbrot argued—contrary to intuition— that you could never find a time during which errors were scattered continuously. Within any burst of errors, no matter how short, there would always be periods of completely error-free transmission. Furthermore, he discovered a consistent geometric relationship between the bursts of errors and the spaces of clean transmission. On scales of an hour or a second, the proportion of error-free periods to error-ridden periods remained constant. (Once, to Mandelbrot's horror, a batch of data seemed to contradict his scheme—but it turned out that the engineers had failed to record the most extreme cases, on the assumption that they were irrelevant.)

Engineers had no framework for understanding Mandelbrot's description, but mathematicians did. In effect, Mandelbrot was duplicating an abstract construction known as the Cantor set, after the nineteenth-century mathematician Georg Cantor. To make a Cantor set, you start with the interval of numbers from zero to one, represented by a line segment. Then you remove the middle third. That leaves two segments, and you remove the middle third of each (from one-ninth to two-ninths and from seven-ninths to eight-ninths). That leaves four segments, and you remove the middle third of each—and so on to infinity. What remains? A strange "dust" of points, arranged in clusters, infinitely many yet infinitely sparse. Mandelbrot was thinking of transmission errors as a Cantor set arranged in time.

This highly abstract description had practical weight for scientists trying to decide between different strategies of controlling error. In particular, it meant that, instead of trying to increase signal strength to drown out more and more noise, engineers should settle for a modest signal, accept the inevitability of errors and use a strategy of redundancy to catch and correct them. Mandelbrot also changed the way IBM's engineers thought about the cause of noise. Bursts of errors had always sent the engineers looking for a man sticking a screwdriver somewhere. But Mandelbrot's scaling patterns suggested that the noise would never be explained on the basis of specific local events.

Mandelbrot turned to other data, drawn from the world's rivers. Egyptians have kept records of the height of the Nile for millennia. It is a matter of more than passing concern. The Nile suffers unusually great variation, flooding heavily in some years and subsiding in others. Mandelbrot classified the variation in terms of two kinds of effects, common in economics as well, which he called the Noah and Joseph Effects.

The Noah Effect means discontinuity: when a quantity changes, it can change almost arbitrarily fast. Economists traditionally

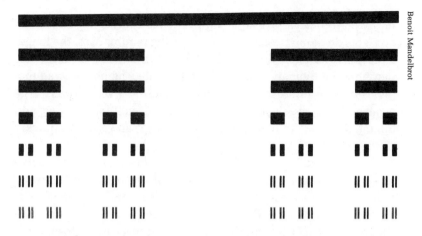

Benoit Mandelbrot

THE CANTOR DUST. Begin with a line; remove the middle third; then remove the middle third of the remaining segments; and so on. The Cantor set is the dust of points that remains. They are infinitely many, but their total length is 0.

The paradoxical qualities of such constructions disturbed nineteenth-century mathematicians, but Mandelbrot saw the Cantor set as a model for the occurrence of errors in an electronic transmission line. Engineers saw periods of error-free transmission, mixed with periods when errors would come in bursts. Looked at more closely, the bursts, too, contained error-free periods within them. And so on—it was an example of fractal time. At every time scale, from hours to seconds, Mandelbrot discovered that the relationship of errors to clean transmission remained constant. Such dusts, he contended, are indispensable in modeling intermittency.

imagined that prices change smoothly—rapidly or slowly, as the case may be, but smoothly in the sense that they pass through all the intervening levels on their way from one point to another. That image of motion was borrowed from physics, like much of the mathematics applied to economics. But it was wrong. Prices can change in instantaneous jumps, as swiftly as a piece of news can flash across a teletype wire and a thousand brokers can change their minds. A stock market strategy was doomed to fail, Mandelbrot argued, if it assumed that a stock would have to sell for $50 at some point on its way down from $60 to $10.

The Joseph Effect means persistence. *There came seven years of great plenty throughout the land of Egypt. And there shall arise after them seven years of famine.* If the Biblical legend meant to imply periodicity, it was oversimplified, of course. But floods and droughts do persist. Despite an underlying randomness, the longer a place has suffered drought, the likelier it is to suffer more. Furthermore, mathematical analysis of the Nile's height showed that persistence applied over centuries as well as over decades. The Noah and Joseph Effects push in different directions, but they add up to this: trends in nature are real, but they can vanish as quickly as they come.

Discontinuity, bursts of noise, Cantor dusts—phenomena like these had no place in the geometries of the past two thousand years. The shapes of classical geometry are lines and planes, circles and spheres, triangles and cones. They represent a powerful abstraction of reality, and they inspired a powerful philosophy of Platonic harmony. Euclid made of them a geometry that lasted two millennia, the only geometry still that most people ever learn. Artists found an ideal beauty in them, Ptolemaic astronomers built a theory of the universe out of them. But for understanding complexity, they turn out to be the wrong kind of abstraction.

Clouds are not spheres, Mandelbrot is fond of saying. Mountains are not cones. Lightning does not travel in a straight line. The new geometry mirrors a universe that is rough, not rounded, scabrous, not smooth. It is a geometry of the pitted, pocked, and broken up, the twisted, tangled, and intertwined. The understanding of nature's complexity awaited a suspicion that the complexity was not just random, not just accident. It required a faith that the interesting feature of a lightning bolt's path, for example, was not its direction, but rather the distribution of zigs and zags. Mandelbrot's work made a claim about the world, and the claim was that such odd shapes carry meaning. The pits and tangles are more than blemishes distorting the classic shapes of Euclidian geometry. They are often the keys to the essence of a thing.

What is the essence of a coastline, for example? Mandelbrot asked this question in a paper that became a turning point for his thinking: "How Long Is the Coast of Britain?"

Mandelbrot had come across the coastline question in an obscure posthumous article by an English scientist, Lewis F. Richardson, who groped with a surprising number of the issues that later became part of chaos. He wrote about numerical weather prediction in the 1920s, studied fluid turbulence by throwing a sack of white parsnips into the Cape Cod Canal, and asked in a 1926 paper, "Does the Wind Possess a Velocity?" ("The question, at first sight foolish, improves on acquaintance," he wrote.) Wondering about coastlines and wiggly national borders, Richardson checked encyclopedias in Spain and Portugal, Belgium and the Netherlands and discovered discrepancies of twenty percent in the estimated lengths of their common frontiers.

Mandelbrot's analysis of this question struck listeners as either painfully obvious or absurdly false. He found that most people answered the question in one of two ways: "I don't know, it's not my field," or "I don't know, but I'll look it up in the encyclopedia."

In fact, he argued, any coastline is—in a sense—infinitely

Richard F. Voss

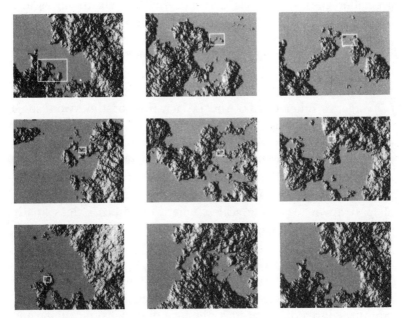

A FRACTAL SHORE. A computer-generated coastline: the details are random, but the fractal dimension is constant, so the degree of roughness or irregularity looks the same no matter how much the image is magnified.

long. In another sense, the answer depends on the length of your ruler. Consider one plausible method of measuring. A surveyor takes a set of dividers, opens them to a length of one yard, and walks them along the coastline. The resulting number of yards is just an approximation of the true length, because the dividers skip over twists and turns smaller than one yard, but the surveyor writes the number down anyway. Then he sets the dividers to a smaller length—say, one foot—and repeats the process. He arrives at a somewhat greater length, because the dividers will capture more of the detail and it will take more than three one-foot steps to cover the distance previously covered by a one-yard step. He writes this new number down, sets the dividers at four inches, and starts again. This mental experiment, using imaginary dividers, is a way of quantifying the effect of observing an object from different distances, at different scales. An observer trying to estimate the length of England's coastline from a satellite will make a smaller guess than an observer trying to walk its coves and beaches, who will make a smaller guess in turn than a snail negotiating every pebble.

Common sense suggests that, although these estimates will continue to get larger, they will approach some particular final value, the true length of the coastline. The measurements should converge, in other words. And in fact, if a coastline were some Euclidean shape, such as a circle, this method of summing finer and finer straight-line distances would indeed converge. But Mandelbrot found that as the scale of measurement becomes smaller, the measured length of a coastline rises without limit, bays and peninsulas revealing ever-smaller subbays and subpeninsulas—at least down to atomic scales, where the process does finally come to an end. Perhaps.

SINCE EUCLIDEAN MEASUREMENTS—length, depth, thickness—failed to capture the essence of irregular shapes, Mandelbrot turned to a different idea, the idea of dimension. Dimension is a quality with a much richer life for scientists than for nonscientists. We live in a three-dimensional world, meaning that we need three numbers to specify a point: for example, longitude, latitude, and altitude. The three dimensions are imagined as directions at right angles to one another. This is still the legacy of Euclidean geometry, where space has three dimensions, a plane has two, a line has one, and a point has zero.

The process of abstraction that allowed Euclid to conceive of one- or two-dimensional objects spills over easily into our use of everyday objects. A road map, for all practical purposes, is a quintessentially two-dimensional thing, a piece of a plane. It uses its two dimensions to carry information of a precisely two-dimensional kind. In reality, of course, road maps are as three-dimensional as everything else, but their thickness is so slight (and so irrelevant to their purpose) that it can be forgotten. Effectively, a road map remains two-dimensional, even when it is folded up. In the same way, a thread is effectively one-dimensional and a particle has effectively no dimension at all.

Then what is the dimension of a ball of twine? Mandelbrot answered, It depends on your point of view. From a great distance, the ball is no more than a point, with zero dimensions. From closer, the ball is seen to fill spherical space, taking up three dimensions. From closer still, the twine comes into view, and the object becomes effectively one-dimensional, though the one dimension is certainly tangled up around itself in a way that makes use of three-dimensional space. The notion of how many numbers it takes to specify a point remains useful. From far away, it takes none—the point is all there is. From closer, it takes three. From closer still, one is enough—any given position along the length of twine is unique, whether the twine is stretched out or tangled up in a ball.

And on toward microscopic perspectives: twine turns to three-dimensional columns, the columns resolve themselves into one-dimensional fibers, the solid material dissolves into zero-dimensional points. Mandelbrot appealed, unmathematically, to relativity: "The notion that a numerical result should depend on the relation of object to observer is in the spirit of physics in this century and is even an exemplary illustration of it."

But philosophy aside, the effective dimension of an object does turn out to be different from its mundane three dimensions. A weakness in Mandelbrot's verbal argument seemed to be its reliance on vague notions, "from far away" and "a little closer." What about in between? Surely there was no clear boundary at which a ball of twine changes from a three-dimensional to a one-dimensional object. Yet, far from being a weakness, the ill-defined nature of these transitions led to a new idea about the problem of dimensions.

Mandelbrot moved beyond dimensions 0, 1, 2, 3 . . . to a seeming impossibility: fractional dimensions. The notion is a conceptual high-wire act. For nonmathematicians it requires a willing suspension of disbelief. Yet it proves extraordinarily powerful.

Fractional dimension becomes a way of measuring qualities that otherwise have no clear definition: the degree of roughness or brokenness or irregularity in an object. A twisting coastline, for example, despite its immeasurability in terms of *length,* nevertheless has a certain characteristic degree of roughness. Mandelbrot specified ways of calculating the fractional dimension of real objects, given some technique of constructing a shape or given some data, and he allowed his geometry to make a claim about the irregular patterns he had studied in nature. The claim was that the degree of irregularity remains constant over different scales. Surprisingly often, the claim turns out to be true. Over and over again, the world displays a regular irregularity.

One wintry afternoon in 1975, aware of the parallel currents emerging in physics, preparing his first major work for publication in book form, Mandelbrot decided he needed a name for his shapes, his dimensions, and his geometry. His son was home from school, and Mandelbrot found himself thumbing through the boy's Latin dictionary. He came across the adjective *fractus,* from the verb *frangere,* to break. The resonance of the main English cognates—*fracture* and *fraction*—seemed appropriate. Mandelbrot created the word (noun and adjective, English and French) *fractal.*

IN THE MIND'S EYE, a fractal is a way of seeing infinity.

Imagine a triangle, each of its sides one foot long. Now imagine a certain transformation—a particular, well-defined, easily repeated set of rules. Take the middle one-third of each side and attach a new triangle, identical in shape but one-third the size.

The result is a star of David. Instead of three one-foot segments, the outline of this shape is now twelve four-inch segments. Instead of three points, there are six.

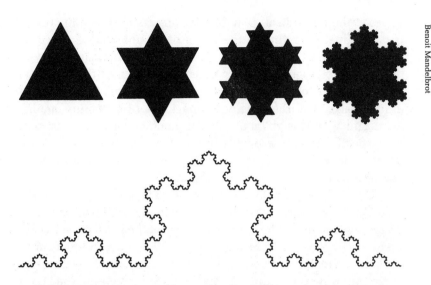

THE KOCH SNOWFLAKE. "A rough but vigorous model of a coastline," in Mandelbrot's words. To construct a Koch curve, begin with a triangle with sides of length 1. At the middle of each side, add a new triangle one-third the size; and so on. The length of the boundary is 3 × 4/3 × 4/3 × 4/3 . . .—infinity. Yet the area remains less than the area of a circle drawn around the original triangle. Thus an infinitely long line surrounds a finite area.

Now take each of the twelve sides and repeat the transformation, attaching a smaller triangle onto the middle third. Now again, and so on to infinity. The outline becomes more and more detailed, just as a Cantor set becomes more and more sparse. It resembles a sort of ideal snowflake. It is known as a Koch curve— a curve being any connected line, whether straight or round— after Helge von Koch, the Swedish mathematician who first described it in 1904.

On reflection, it becomes apparent that the Koch curve has some interesting features. For one thing, it is a continuous loop, never intersecting itself, because the new triangles on each side are always small enough to avoid bumping into each other. Each transformation adds a little area to the inside of the curve, but the total area remains finite, not much bigger than the original triangle, in fact. If you drew a circle around the original triangle, the Koch curve would never extend beyond it.

Yet the curve itself is infinitely long, as long as a Euclidean straight line extending to the edges of an unbounded universe. Just as the first transformation replaces a one-foot segment with four four-inch segments, every transformation multiplies the total length by four-thirds. This paradoxical result, infinite length in a finite space, disturbed many of the turn-of-the-century mathematicians who thought about it. The Koch curve was monstrous, disrespectful to all reasonable intuition about shapes and—it almost went without saying—pathologically unlike anything to be found in nature.

Under the circumstances, their work made little impact at the time, but a few equally perverse mathematicians imagined other shapes with some of the bizarre qualities of the Koch curve. There were Peano curves. There were Sierpiński carpets and Sierpiński gaskets. To make a carpet, start with a square, divide it three-by-three into nine equal squares, and remove the central one. Then repeat the operation on the eight remaining squares, putting a square hole in the center of each. The gasket is the same but with equilateral triangles instead of squares; it has the hard-to-imagine property that any arbitrary point is a branching point, a fork in the structure. Hard to imagine, that is, until you have thought about the Eiffel Tower, a good three-dimensional approximation, its beams and trusses and girders branching into a lattice of ever-thinner members, a shimmering network of fine detail. Eiffel, of course, could not carry the scheme to infinity, but he appreciated the subtle engineering point that allowed him to remove weight without also removing structural strength.

The mind cannot visualize the whole infinite self-embedding of complexity. But to someone with a geometer's way of thinking about form, this kind of repetition of structure on finer and finer scales can open a whole world. Exploring these shapes, pressing one's mental fingers into the rubbery edges of their possibilities, was a kind of playing, and Mandelbrot took a childlike delight in seeing variations that no one had seen or understood before. When they had no names, he named them: ropes and sheets, sponges and foams, curds and gaskets.

Fractional dimension proved to be precisely the right yardstick. In a sense, the degree of irregularity corresponded to the efficiency of the object in taking up space. A simple, Euclidean,

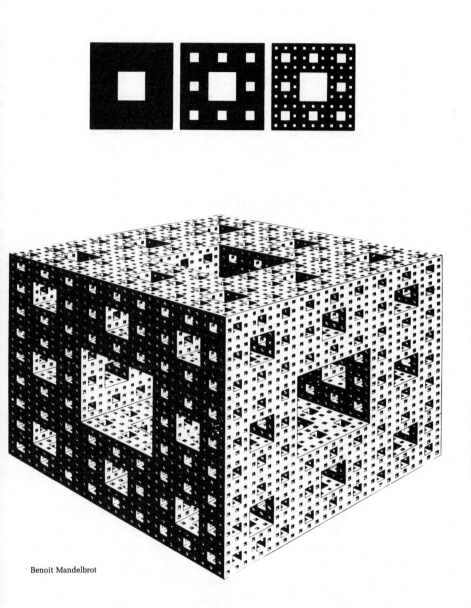

Benoit Mandelbrot

CONSTRUCTING WITH HOLES. A few mathematicians in the early twentieth century conceived monstrous-seeming objects made by the technique of adding or removing infinitely many parts. One such shape is the Sierpiński carpet, constructed by cutting the center one-ninth of a square; then cutting out the centers of the eight smaller squares that remain; and so on. The three-dimensional analogue is the Menger sponge, a solid-looking lattice that has an infinite surface area, yet zero volume.

one-dimensional line fills no space at all. But the outline of the Koch curve, with infinite length crowding into finite area, does fill space. It is more than a line, yet less than a plane. It is greater than one-dimensional, yet less than a two-dimensional form. Using techniques originated by mathematicians early in the century and then all but forgotten, Mandelbrot could characterize the fractional dimension precisely. For the Koch curve, the infinitely extended multiplication by four-thirds gives a dimension of 1.2618.

In pursuing this path, Mandelbrot had two great advantages over the few other mathematicians who had thought about such shapes. One was his access to the computing resources that go with the name of IBM. Here was another task ideally suited to the computer's particular form of high-speed idiocy. Just as meteorologists needed to perform the same few calculations at millions of neighboring points in the atmosphere, Mandelbrot needed to perform an easily programmed transformation again and again and again and again. Ingenuity could conceive of transformations. Computers could draw them—sometimes with unexpected results. The early twentieth-century mathematicians quickly reached a barrier of hard calculation, like the barrier faced by early protobiologists without microscopes. In looking into a universe of finer and finer detail, the imagination can carry one only so far.

In Mandelbrot's words: "There was a long hiatus of a hundred years where drawing did not play any role in mathematics because hand and pencil and ruler were exhausted. They were well understood and no longer in the forefront. And the computer did not exist.

"When I came in this game, there was a total absence of intuition. One had to create an intuition from scratch. Intuition as it was trained by the usual tools—the hand, the pencil, and the ruler—found these shapes quite monstrous and pathological. The old intuition was misleading. The first pictures were to me quite a surprise; then I would recognize some pictures from previous pictures, and so on.

"Intuition is not something that is given. I've trained my intuition to accept as obvious shapes which were initially rejected as absurd, and I find everyone else can do the same."

Mandelbrot's other advantage was the picture of reality he had begun forming in his encounters with cotton prices, with electronic transmission noise, and with river floods. The picture was beginning to come into focus now. His studies of irregular patterns in natural processes and his exploration of infinitely complex shapes had an intellectual intersection: a quality of *self-similarity*. Above all, fractal meant self-similar.

Self-similarity is symmetry across scale. It implies recursion, pattern inside of pattern. Mandelbrot's price charts and river charts displayed self-similarity, because not only did they produce detail at finer and finer scales, they also produced detail with certain constant measurements. Monstrous shapes like the Koch curve display self-similarity because they look exactly the same even under high magnification. The self-similarity is built into the technique of constructing the curves—the same transformation is repeated at smaller and smaller scales. Self-similarity is an easily recognizable quality. Its images are everywhere in the culture: in the infinitely deep reflection of a person standing between two mirrors, or in the cartoon notion of a fish eating a smaller fish eating a smaller fish eating a smaller fish. Mandelbrot likes to quote Jonathan Swift: "So, Nat'ralists observe, a Flea/Hath smaller Fleas that on him prey,/And these have smaller Fleas to bite 'em,/ And so proceed ad infinitum."

IN THE NORTHEASTERN United States, the best place to study earthquakes is the Lamont-Doherty Geophysical Observatory, a group of unprepossessing buildings hidden in the woods of southern New York State, just west of the Hudson River. Lamont-Doherty is where Christopher Scholz, a Columbia University professor specializing in the form and structure of the solid earth, first started thinking about fractals.

While mathematicians and theoretical physicists disregarded Mandelbrot's work, Scholz was precisely the kind of pragmatic, working scientist most ready to pick up the tools of fractal geometry. He had stumbled across Benoit Mandelbrot's name in the 1960s, when Mandelbrot was working in economics and Scholz was an M.I.T. graduate student spending a great deal of time on a stubborn question about earthquakes. It had been well known for twenty years that the distribution of large and small earthquakes followed a particular mathematical pattern, precisely the same scaling pattern that seemed to govern the distribution of personal incomes in a free-market economy. This distribution was observed everywhere on earth, wherever earthquakes were counted and measured. Considering how irregular and unpredictable earthquakes were otherwise, it was worthwhile to ask what sort of physical processes might explain this regularity. Or so it seemed to Scholz. Most seismologists had been content to note the fact and move on.

Scholz remembered Mandelbrot's name, and in 1978 he bought a profusely illustrated, bizarrely erudite, equation-studded book called *Fractals: Form, Chance and Dimension*. It was as if Mandelbrot had collected in one rambling volume everything he knew or suspected about the universe. Within a few years this book and its expanded and refined replacement, *The Fractal Geometry of Nature*, had sold more copies than any other book of high mathematics. Its style was abstruse and exasperating, by turns witty, literary, and opaque. Mandelbrot himself called it "a manifesto and a casebook."

Like a few counterparts in a handful of other fields, particularly scientists who worked on the material parts of nature, Scholz spent several years trying to figure out what to do with this book. It was far from obvious. *Fractals* was, as Scholz put it, "not a how-to book but a gee-whiz book." Scholz, however, happened to care deeply about surfaces, and surfaces were everywhere in this book. He found that he could not stop thinking about the promise of Mandelbrot's ideas. He began to work out a way of using fractals to describe, classify, and measure the pieces of his scientific world.

He soon realized that he was not alone, although it was several more years before fractals conferences and seminars began multiplying. The unifying ideas of fractal geometry brought together scientists who thought their own observations were idiosyncratic and who had no systematic way of understanding them. The insights of fractal geometry helped scientists who study the way things meld together, the way they branch apart, or the way they shatter. It is a method of looking at materials—the microscopically jagged surfaces of metals, the tiny holes and channels of porous oil-bearing rock, the fragmented landscapes of an earthquake zone.

As Scholz saw it, it was the business of geophysicists to describe the surface of the earth, the surface whose intersection with the flat oceans makes coastlines. Within the top of the solid earth are surfaces of another kind, surfaces of cracks. Faults and fractures so dominate the structure of the earth's surface that they become the key to any good description, more important on balance than the material they run through. The fractures crisscross the earth's surface in three dimensions, creating what Scholz whimsically called the "schizosphere." They control the flow of fluid through the ground—the flow of water, the flow of oil, and the flow of natural gas. They control the behavior of earthquakes. Understanding surfaces was paramount, yet Scholz believed that his profession was in a quandary. In truth, no framework existed.

Geophysicists looked at surfaces the way anyone would, as shapes. A surface might be flat. Or it might have a particular shape. You could look at the outline of a Volkswagen Beetle, for example, and draw that surface as a curve. The curve would be measurable in familiar Euclidean ways. You could fit an equation to it. But in Scholz's description, you would only be looking at that surface through a narrow spectral band. It would be like looking at the universe through a red filter—you see what is happening at that particular wavelength of light, but you miss everything happening at the wavelengths of other colors, not to mention that vast range of activity at parts of the spectrum corresponding to infrared radiation or radio waves. The spectrum, in this analogy, corresponds to scale. To think of the surface of a Volkswagen in terms of its Euclidean shape is to see it only on the scale of an observer ten meters or one hundred meters away. What about an observer one kilometer away, or one hundred kilometers? What about an observer one millimeter away, or one micron?

Imagine tracing the surface of the earth as it would look from a distance of one hundred kilometers out in space. The line goes up and down over trees and hillocks, buildings and—in a parking lot somewhere—a Volkswagen. On that scale, the surface is just a bump among many other bumps, a bit of randomness.

Or imagine looking at the Volkswagen from closer and closer, zooming in with magnifying glass and microscope. At first the surface seems to get smoother, as the roundness of bumpers and hood passes out of view. But then the microscopic surface of the steel turns out to be bumpy itself, in an apparently random way. It seems chaotic.

Scholz found that fractal geometry provided a powerful way of describing the particular bumpiness of the earth's surface, and metallurgists found the same for the surfaces of different kinds of steel. The fractal dimension of a metal's surface, for example, often provides information that corresponds to the metal's strength. And the fractal dimension of the earth's surface provides clues to its important qualities as well. Scholz thought about a classic geological formation, a talus slope on a mountainside. From a distance it is a Euclidean shape, dimension two. As a geologist approaches, though, he finds himself walking not so much on it as in it—the talus has resolved itself into boulders the size of cars. Its effective dimension has become about 2.7, because the rock surfaces hook over and wrap around and nearly fill three-dimensional space, like the surface of a sponge.

Fractal descriptions found immediate application in a series of problems connected to the properties of surfaces in contact with one another. The contact between tire treads and concrete is such a problem. So is contact in machine joints, or electrical contact. Contacts between surfaces have properties quite independent of the materials involved. They are properties that turn out to depend on the fractal quality of the bumps upon bumps upon bumps. One simple but powerful consequence of the fractal geometry of surfaces is that surfaces in contact do not touch everywhere. The bumpiness at all scales prevents that. Even in rock under enormous pressure, at some sufficiently small scale it becomes clear that gaps remain, allowing fluid to flow. To Scholz, it is the Humpty-Dumpty Effect. It is why two pieces of a broken teacup can never be rejoined, even though they appear to fit together at some gross scale. At a smaller scale, irregular bumps are failing to coincide.

Scholz became known in his field as one of a few people taking up fractal techniques. He knew that some of his colleagues viewed this small group as freaks. If he used the word fractal in the title of a paper, he felt that he was regarded either as being admirably current or not-so-admirably on a bandwagon. Even the writing of papers forced difficult decisions, between writing for a small audience of fractal aficionados or writing for a broader geophysical audience that would need explanations of the basic concepts. Still, Scholz considered the tools of fractal geometry indispensable.

"It's a single model that allows us to cope with the range of changing dimensions of the earth," he said. "It gives you mathematical and geometrical tools to describe and make predictions. Once you get over the hump, and you understand the paradigm, you can start actually measuring things and thinking about things in a new way. You see them differently. You have a new vision. It's not the same as the old vision at all—it's much broader."

HOW BIG IS IT? How long does it last? These are the most basic questions a scientist can ask about a thing. They are so basic to the way people conceptualize the world that it is not easy to see that they imply a certain bias. They suggest that size and duration, qualities that depend on scale, are qualities with meaning, qualities that can help describe an object or classify it. When a biologist describes a human being, or a physicist describes a quark, how big and how long are indeed appropriate questions. In their gross physical structure, animals are very much tied to a particular scale. Imagine a human being scaled up to twice its size, keeping all proportions the same, and you imagine a structure whose bones will collapse under its weight. Scale is important.

The physics of earthquake behavior is mostly independent of scale. A large earthquake is just a scaled-up version of a small earthquake. That distinguishes earthquakes from animals, for example—a ten-inch animal must be structured quite differently from a one-inch animal, and a hundred-inch animal needs a different architecture still, if its bones are not to snap under the increased mass. Clouds, on the other hand, are scaling phenomena like earthquakes. Their characteristic irregularity—describable in terms of fractal dimension—changes not at all as they are observed on different scales. That is why air travelers lose all perspective on how far away a cloud is. Without help from cues such as haziness, a cloud twenty feet away can be indistinguishable from two thousand feet away. Indeed, analysis of satellite pictures has shown an invariant fractal dimension in clouds observed from hundreds of miles away.

It is hard to break the habit of thinking of things in terms of how big they are and how long they last. But the claim of fractal geometry is that, for some elements of nature, looking for a characteristic scale becomes a distraction. *Hurricane.* By definition, it is a storm of a certain size. But the definition is imposed by people on nature. In reality, atmospheric scientists are realizing that tumult in the air forms a continuum, from the gusty swirling of litter on a city street corner to the vast cyclonic systems visible from space. Categories mislead. The ends of the continuum are of a piece with the middle.

It happens that the equations of fluid flow are in many contexts dimensionless, meaning that they apply without regard to scale. Scaled-down airplane wings and ship propellers can be tested in wind tunnels and laboratory basins. And, with some limitations, small storms act like large storms.

Blood vessels, from aorta to capillaries, form another kind of continuum. They branch and divide and branch again until they become so narrow that blood cells are forced to slide through single file. The nature of their branching is fractal. Their structure resembles one of the monstrous imaginary objects conceived by Mandelbrot's turn-of-the-century mathematicians. As a matter of physiological necessity, blood vessels must perform a bit of dimensional magic. Just as the Koch curve, for example, squeezes a line of infinite length into a small area, the circulatory system must squeeze a huge surface area into a limited volume. In terms of the body's resources, blood is expensive and space is at a premium. The fractal structure nature has devised works so efficiently that, in most tissue, no cell is ever more than three or four cells away from a blood vessel. Yet the vessels and blood take up little space, no more than about five percent of the body. It is, as Mandelbrot put it, the Merchant of Venice Syndrome—not only can't you take a pound of flesh without spilling blood, you can't take a milligram.

This exquisite structure—actually, two intertwining trees of veins and arteries—is far from exceptional. The body is filled with such complexity. In the digestive tract, tissue reveals undulations within undulations. The lungs, too, need to pack the greatest possible surface into the smallest space. An animal's ability to absorb oxygen is roughly proportional to the surface area of its lungs. Typical human lungs pack in a surface bigger than a tennis court. As an added complication, the labyrinth of windpipes must merge efficiently with the arteries and veins.

Every medical student knows that lungs are designed to accommodate a huge surface area. But anatomists are trained to look at one scale at a time—for example, at the millions of alveoli, microscopic sacs, that end the sequence of branching pipes. The language of anatomy tends to obscure the unity *across* scales. The fractal approach, by contrast, embraces the whole structure in terms of the branching that produces it, branching that behaves consistently from large scales to small. Anatomists study the vasculatory system by classifying blood vessels into categories based on size—arteries and arterioles, veins and venules. For some purposes, those categories prove useful. But for others they mislead. Sometimes the textbook approach seems to dance around the truth: "In the gradual transition from one type of artery to another it is sometimes difficult to classify the intermediate region. Some arteries of intermediate caliber have walls that suggest larger arteries, while some large arteries have walls like those of medium-sized arteries. The transitional regions . . . are often designated arteries of mixed type."

Not immediately, but a decade after Mandelbrot published his physiological speculations, some theoretical biologists began to find fractal organization controlling structures all through the body. The standard "exponential" description of bronchial branching proved to be quite wrong; a fractal description turned out to fit the data. The urinary collecting system proved fractal. The biliary duct in the liver. The network of special fibers in the heart that carry pulses of electric current to the contracting muscles. The last structure, known to heart specialists as the His-Purkinje network, inspired a particularly important line of research. Considerable work on healthy and abnormal hearts turned out to hinge on the details of how the muscle cells of the left and right pumping chambers all manage to coordinate their timing. Several chaos-minded cardiologists found that the frequency spectrum of heartbeat timing, like earthquakes and economic phenomena, followed fractal laws, and they argued that one key to understanding heartbeat timing was the fractal organization of the His-Purkinje network, a labyrinth of branching pathways organized to be self-similar on smaller and smaller scales.

How did nature manage to evolve such complicated architecture? Mandelbrot's point is that the complications exist only in the context of traditional Euclidean geometry. As fractals, branching structures can be described with transparent simplicity, with just a few bits of information. Perhaps the simple transformations that gave rise to the shapes devised by Koch, Peano, and Sierpiński have their analogue in the coded instructions of an organism's genes. DNA surely cannot specify the vast number of bronchi, bronchioles, and alveoli or the particular spatial structure of the resulting tree, but it can specify a repeating process of bifurcation and development. Such processes suit nature's purposes. When E. I. DuPont de Nemours & Company and the United States Army finally began to produce a synthetic match for goose down, it was by finally realizing that the phenomenal air-trapping ability of the natural product came from the fractal nodes and branches of down's key protein, keratin. Mandelbrot glided matter-of-factly from pulmonary and vascular trees to real botanical trees, trees that need to capture sun and resist wind, with fractal branches and fractal leaves. And theoretical biologists began to speculate that fractal scaling was not just common but universal in morphogenesis. They argued that understanding how such patterns were encoded and processed had become a major challenge to biology.

"I STARTED LOOKING in the trash cans of science for such phenomena, because I suspected that what I was observing was not an exception but perhaps very widespread. I attended lectures and looked in unfashionable periodicals, most of them of little or no yield, but once in a while finding some interesting things. In a way it was a naturalist's approach, not a theoretician's approach. But my gamble paid off."

Having consolidated a life's collection of ideas about nature and mathematical history into one book, Mandelbrot found an unaccustomed measure of academic success. He became a fixture of the scientific lecture circuit, with his indispensable trays of color slides and his wispy white hair. He began to win prizes and other professional honors, and his name became as well known to the nonscientific public as any mathematician's. In part that was because of the aesthetic appeal of his fractal pictures; in part because the many thousands of hobbyists with microcomputers could begin exploring his world themselves. In part it was because he put himself forward. His name appeared on a little list compiled by the Harvard historian of science I. Bernard Cohen. Cohen had scoured the annals of discovery for years, looking for scientists who had declared their own work to be "revolutions." All told, he found just sixteen. Robert Symmer, a Scots contemporary of Benjamin Franklin whose ideas about electricity were indeed radical, but wrong. Jean-Paul Marat, known today only for his bloody contribution to the French Revolution. Von Liebig. Hamilton. Charles Darwin, of course. Virchow. Cantor. Einstein. Minkowski. Von Laue. Alfred Wegener—continental drift. Compton. Just. James Watson—the structure of DNA. And Benoit Mandelbrot.

To pure mathematicians, however, Mandelbrot remained an outsider, contending as bitterly as ever with the politics of science. At the height of his success, he was reviled by some colleagues, who thought he was unnaturally obsessed with his place in history. They said he hectored them about giving due credit. Unquestionably, in his years as a professional heretic he honed an appreciation for the tactics as well as the substance of scientific achievement. Sometimes when articles appeared using ideas from fractal geometry he would call or write the authors to complain that no reference was made to him or his book.

His admirers found his ego easy to forgive, considering the difficulties he had overcome in getting recognition for his work. "Of course, he is a bit of a megalomaniac, he has this incredible ego, but it's beautiful stuff he does, so most people let him get away with it," one said. In the words of another: "He had so many difficulties with his fellow mathematicians that simply in order to survive he had to develop this strategy of boosting his own ego. If he hadn't done that, if he hadn't been so convinced that he had the right visions, then he would never have succeeded."

The business of taking and giving credit can become obsessive in science. Mandelbrot did plenty of both. His book rings with the first person: I claim . . . I conceived and developed . . . and implemented . . . I have confirmed . . . I show . . . I coined . . . In my travels through newly opened or newly settled territory, I was often moved to exert the right of naming its landmarks.

Many scientists failed to appreciate this kind of style. Nor were they mollified that Mandelbrot was equally copious with his references to predecessors, some thoroughly obscure. (And all, as his detractors noted, quite safely deceased.) They thought it was just his way of trying to position himself squarely in the center, setting himself up like the Pope, casting his benedictions from one side of the field to the other. They fought back. Scientists could hardly avoid the word *fractal*, but if they wanted to avoid Mandelbrot's name they could speak of fractional dimension as *Hausdorff-Besicovitch dimension*. They also—particularly mathematicians—resented the way he moved in and out of different disciplines, making his claims and conjectures and leaving the real work of proving them to others.

It was a legitimate question. If one scientist announces that a thing is probably true, and another demonstrates it with rigor, which one has done more to advance science? Is the making of a conjecture an act of discovery? Or is it just a cold-blooded staking of a claim? Mathematicians have always faced such issues, but the debate became more intense as computers began to play their new role. Those who used computers to conduct experiments became more like laboratory scientists, playing by rules that allowed discovery without the usual theorem-proof, theorem-proof of the standard mathematics paper.

Mandelbrot's book was wide-ranging and stuffed with the minutiae of mathematical history. Wherever chaos led, Mandelbrot had some basis to claim that he had been there first. Little did it matter that most readers found his references obscure or even useless. They had to acknowledge his extraordinary intuition for the direction of advances in fields he had never actually studied, from seismology to physiology. It was sometimes uncanny, and sometimes irritating. Even an admirer would cry with exasperation, "Mandelbrot didn't have *everybody's* thoughts before they did."

It hardly matters. The face of genius need not always wear an Einstein's saintlike mien. Yet for decades, Mandelbrot believes, he had to play games with his work. He had to couch original ideas in terms that would not give offense. He had to delete his visionary-sounding prefaces to get his articles published. When he wrote the first version of his book, published in French in 1975, he felt he was forced to pretend it contained nothing too startling.

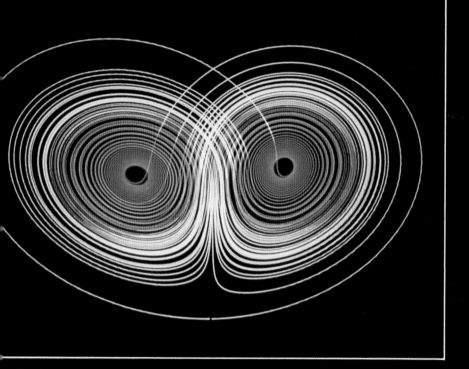

The Lorenz attractor.

The Koch curve.

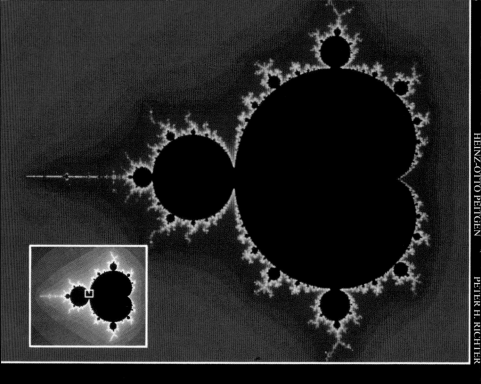

THE MANDELBROT SET. A voyage through finer and finer scales shows the increasing complexity of the set, with its seahorse tails and island molecules resembling the whole set. By the last frame, the level of magnification is about one million in each direction.

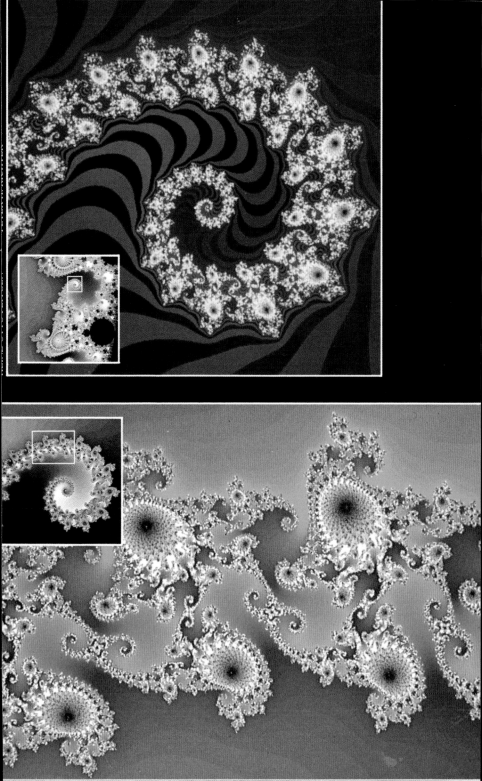

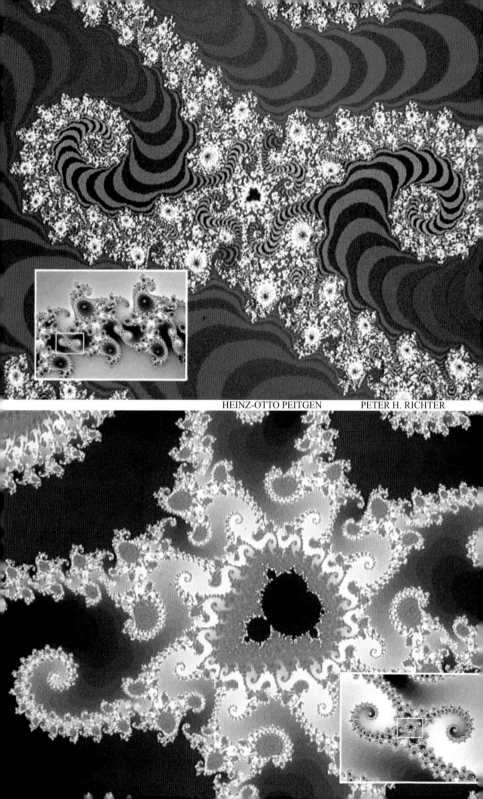

HEINZ-OTTO PEITGEN   PETER H. RICHTER

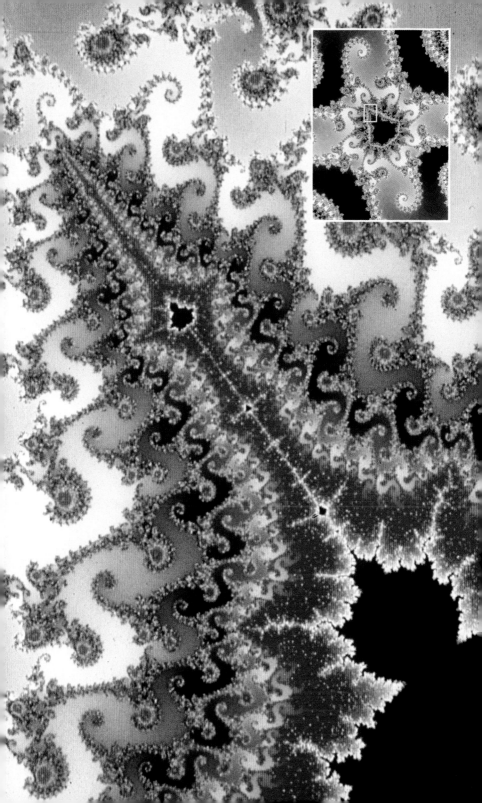

THE COMPLEX BOUNDARIES OF NEWTON'S METHOD. The attracting pull of four points—in the four dark holes—creates "basins of attraction," each a different color, with a complicated fractal boundary. The image represents the way Newton's method for solving equations leads from different starting points to one of four possible solutions (in this case the equation is $x^4 - 1 = 0$).

(Illustrations on next page)

THE GREAT RED SPOT: REAL AND SIMULATED. The Voyager satellite revealed Jupiter's surface as a seething, turbulent fluid, with horizontal bands of east-west flow. The Great Red Spot is seen from above the planet's equator and also in a view looking down on the South Pole.

Computer graphics from Philip Marcus's simulation present the South Pole view. The color shows the direction of spin for particular pieces of fluid: pieces turning counterclockwise are red, and pieces turning clockwise are blue. No matter what the starting configuration, clumps of blue tend to break up, while the red tends to merge into a single spot, stable and coherent amid the surrounding tumult.

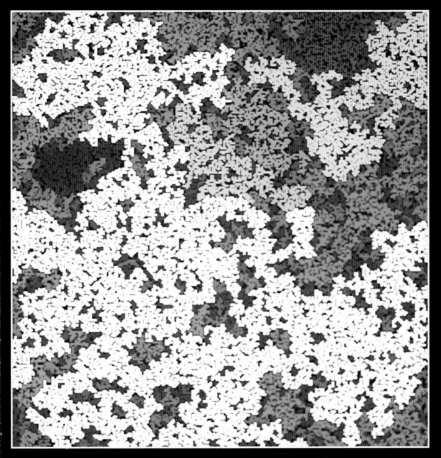

FRACTAL CLUSTERS. A random clustering of particles generated by a computer produces a "percolation network," one of many visual models inspired by fractal geometry. Applied physicists discovered that such models imitate a variety of real-world processes, such as the formation of polymers and the diffusion of oil through fractured rock. Each color in the percolation network represents a grouping that is connected throughout.

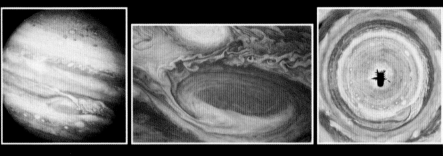

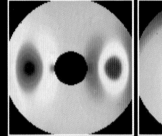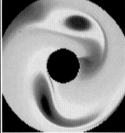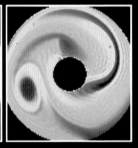

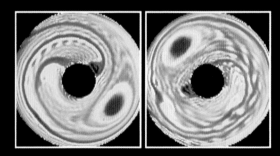

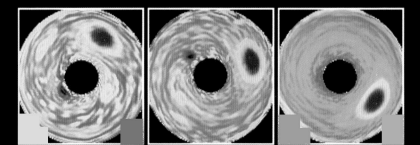

That was why he wrote the latest version explicitly as "a manifesto and a casebook." He was coping with the politics of science.

"The politics affected the style in a sense which I later came to regret. I was saying, 'It's natural to . . . , It's an interesting observation that . . . .' Now, in fact, it was anything but natural, and the interesting observation was in fact the result of very long investigations and search for proof and self-criticism. It had a philosophical and removed attitude which I felt was necessary to get it accepted. The politics was that, if I said I was proposing a radical departure, that would have been the end of the readers' interest.

"Later on, I got back some such statements, people saying, 'It is natural to observe . . .' That was not what I had bargained for."

Looking back, Mandelbrot saw that scientists in various disciplines responded to his approach in sadly predictable stages. The first stage was always the same: Who are you and why are you interested in our field? Second: How does it relate to what we have been doing, and why don't you explain it on the basis of what we know? Third: Are you sure it's standard mathematics? (Yes, I'm sure.) Then why don't we know it? (Because it's standard but very obscure.)

Mathematics differs from physics and other applied sciences in this respect. A branch of physics, once it becomes obsolete or unproductive, tends to be forever part of the past. It may be a historical curiosity, perhaps the source of some inspiration to a modern scientist, but dead physics is usually dead for good reason. Mathematics, by contrast, is full of channels and byways that seem to lead nowhere in one era and become major areas of study in another. The potential application of a piece of pure thought can never be predicted. That is why mathematicians value work in an aesthetic way, seeking elegance and beauty as artists do. It is also why Mandelbrot, in his antiquarian mode, came across so much good mathematics that was ready to be dusted off.

So the fourth stage was this: What do people in these branches of mathematics think about your work? (They don't care, because it doesn't add to the mathematics. In fact, they are surprised that their ideas represent nature.)

In the end, the word *fractal* came to stand for a way of describing, calculating, and thinking about shapes that are irregular and fragmented, jagged and broken-up—shapes from the crystalline curves of snowflakes to the discontinuous dusts of galaxies. A fractal curve implies an organizing structure that lies hidden among the hideous complication of such shapes. High school students could understand fractals and play with them; they were as primary as the elements of Euclid. Simple computer programs to draw fractal pictures made the rounds of personal computer hobbyists.

Mandelbrot found his most enthusiastic acceptance among applied scientists working with oil or rock or metals, particularly in corporate research centers. By the middle of the 1980s, vast numbers of scientists at Exxon's huge research facility, for example, worked on fractal problems. At General Electric, fractals became an organizing principle in the study of polymers and also—though this work was conducted secretly—in problems of nuclear reactor safety. In Hollywood, fractals found their most whimsical application in the creation of phenomenally realistic landscapes, earthly and extraterrestrial, in special effects for movies.

The patterns that people like Robert May and James Yorke discovered in the early 1970s, with their complex boundaries between orderly and chaotic behavior, had unsuspected regularities that could only be described in terms of the relation of large scales to small. The structures that provided the key to nonlinear dynamics proved to be fractal. And on the most immediate practical level, fractal geometry also provided a set of tools that were taken up by physicists, chemists, seismologists, metallurgists, probability theorists and physiologists. These researchers were convinced, and they tried to convince others, that Mandelbrot's new geometry was nature's own.

They made an irrefutable impact on orthodox mathematics and physics as well, but Mandelbrot himself never gained the full respect of those communities. Even so, they had to acknowledge him. One mathematician told friends that he had awakened one night still shaking from a nightmare. In this dream, the mathematician was dead, and suddenly heard the unmistakable voice of God. "You know," He remarked, "there really *was* something to that Mandelbrot."

THE NOTION OF SELF-SIMILARITY strikes ancient chords in our culture. An old strain in Western thought honors the idea. Leibniz imagined that a drop of water contained a whole teeming universe, containing, in turn, water drops and new universes within. "To see the world in a grain of sand," Blake wrote, and often scientists were predisposed to see it. When sperm were first discovered, each was thought to be a homunculus, a human, tiny but fully formed.

But self-similarity withered as a scientific principle, for a good reason. It did not fit the facts. Sperm are not merely scaled-down humans—they are far more interesting than that—and the process of ontogenetic development is far more interesting than mere enlargement. The early sense of self-similarity as an organizing principle came from the limitations on the human experience of scale. How else to imagine the very great and very small, the very fast and very slow, but as extensions of the known?

The myth died hard as the human vision was extended by telescopes and microscopes. The first discoveries were realizations that each change of scale brought new phenomena and new kinds of behavior. For modern particle physicists, the process has never ended. Every new accelerator, with its increase in energy and speed, extends science's field of view to tinier particles and briefer time scales, and every extension seems to bring new information.

At first blush, the idea of consistency on new scales seems to provide less information. In part, that is because a parallel trend in science has been toward reductionism. Scientists break things apart and look at them one at a time. If they want to examine the interaction of subatomic particles, they put two or three together. There is complication enough. The power of self-similarity, though, begins at much greater levels of complexity. It is a matter of looking at the whole.

Although Mandelbrot made the most comprehensive geometric use of it, the return of scaling ideas to science in the 1960s and 1970s became an intellectual current that made itself felt simultaneously in many places. Self-similarity was implicit in

Edward Lorenz's work. It was part of his intuitive understanding of the fine structure of the maps made by his system of equations, a structure he could sense but not see on the computers available in 1963. Scaling also became part of a movement in physics that led, more directly than Mandelbrot's own work, to the discipline known as chaos. Even in distant fields, scientists were beginning to think in terms of theories that used hierarchies of scales, as in evolutionary biology, where it became clear that a full theory would have to recognize patterns of development in genes, in individual organisms, in species, and in families of species, all at once.

Paradoxically, perhaps, the appreciation of scaling phenomena must have come from the same kind of expansion of human vision that had killed the earlier naïve ideas of self-similarity. By the late twentieth century, in ways never before conceivable, images of the incomprehensibly small and the unimaginably large became part of everyone's experience. The culture saw photographs of galaxies and of atoms. No one had to imagine, with Leibniz, what the universe might be like on microscopic or telescopic scales—microscopes and telescopes made those images part of everyday experience. Given the eagerness of the mind to find analogies in experience, new kinds of comparison between large and small were inevitable—and some of them were pro ductive.

Often the scientists drawn to fractal geometry felt emotional parallels between their new mathematical aesthetic and changes in the arts in the second half of the century. They felt that they were drawing some inner enthusiasm from the culture at large. To Mandelbrot the epitome of the Euclidean sensibility outside mathematics was the architecture of the Bauhaus. It might just as well have been the style of painting best exemplified by the color squares of Josef Albers: spare, orderly, linear, reductionist, geometrical. *Geometrical*—the word means what it has meant for thousands of years. Buildings that are called geometrical are composed of simple shapes, straight lines and circles, describable with just a few numbers. The vogue for geometrical architecture and painting came and went. Architects no longer care to build blockish skyscrapers like the Seagram Building in New York, once much hailed and copied. To Mandelbrot and his followers the reason is clear. Simple shapes are inhuman. They fail to resonate with the way nature organizes itself or with the way human perception sees the world. In the words of Gert Eilenberger, a German physicist who took up nonlinear science after specializing in superconductivity: "Why is it that the silhouette of a storm-bent leafless tree against an evening sky in winter is perceived as beautiful, but the corresponding silhouette of any multi-purpose university building is not, in spite of all efforts of the architect? The answer seems to me, even if somewhat speculative, to follow from the new insights into dynamical systems. Our feeling for beauty is inspired by the harmonious arrangement of order and disorder as it occurs in natural objects—in clouds, trees, mountain ranges, or snow crystals. The shapes of all these are dynamical processes jelled into physical forms, and particular combinations of order and disorder are typical for them."

A geometrical shape has a *scale*, a characteristic size. To Mandelbrot, art that satisfies lacks scale, in the sense that it contains important elements at all sizes. Against the Seagram Building, he offers the architecture of the Beaux-Arts, with its sculptures and gargoyles, its quoins and jamb stones, its cartouches decorated with scrollwork, its cornices topped with cheneaux and lined with dentils. A Beaux-Arts paragon like the Paris Opera has no scale because it has every scale. An observer seeing the building from any distance finds some detail that draws the eye. The composition changes as one approaches and new elements of the structure come into play.

Appreciating the harmonious structure of any architecture is one thing; admiring the wildness of nature is quite another. In terms of aesthetic values, the new mathematics of fractal geometry brought hard science in tune with the peculiarly modern feeling for untamed, uncivilized, undomesticated nature. At one time rain forests, deserts, bush, and badlands represented all that society was striving to subdue. If people wanted aesthetic satisfaction from vegetation, they looked at gardens. As John Fowles put it, writing of eighteenth-century England: "The period had no sympathy with unregulated or primordial nature. It was aggressive wilderness, an ugly and all-invasive reminder of the Fall, of man's eternal exile from the Garden of Eden. . . . Even its natural sciences . . . remained essentially hostile to wild nature, seeing it only as something to be tamed, classified, utilised, exploited." By the end of the twentieth century, the culture had changed, and now science was changing with it.

So science found a use after all for the obscure and fanciful cousins of the Cantor set and the Koch curve. At first, these shapes could have served as items of evidence in the divorce proceedings between mathematics and the physical sciences at the turn of the century, the end of a marriage that had been the dominating theme of science since Newton. Mathematicians like Cantor and Koch had delighted in their originality. They thought they were outsmarting nature—when actually they had not yet caught up with nature's creation. The prestigious mainstream of physics, too, turned away from the world of everyday experience. Only later, after Steve Smale brought mathematicians back to dynamical systems, could a physicist say, "We have the astronomers and mathematicians to thank for passing the field on to us, physicists, in a much better shape than we left it to them, 70 years ago."

Yet, despite Smale and despite Mandelbrot, it was to be the physicists after all who made a new science of chaos. Mandelbrot provided an indispensable language and a catalogue of surprising pictures of nature. As Mandelbrot himself acknowledged, his program *described* better than it *explained*. He could list elements of nature along with their fractal dimensions—seacoasts, river networks, tree bark, galaxies—and scientists could use those numbers to make predictions. But physicists wanted to know more. They wanted to know why. There were forms in nature—not visible forms, but shapes embedded in the fabric of motion—waiting to be revealed.

Strange Attractors

Big whorls have little whorls
Which feed on their velocity,
And little whorls have lesser whorls
And so on to viscosity.
                    —LEWIS F. RICHARDSON

TURBULENCE WAS A PROBLEM with pedigree. The great physicists all thought about it, formally or informally. A smooth flow breaks up into whorls and eddies. Wild patterns disrupt the boundary between fluid and solid. Energy drains rapidly from large-scale motions to small. Why? The best ideas came from mathematicians; for most physicists, turbulence was too dangerous to waste time on. It seemed almost unknowable. There was a story about the quantum theorist Werner Heisenberg, on his deathbed, declaring that he will have two questions for God: why relativity, and why turbulence. Heisenberg says, "I really think He may have an answer to the first question."

Theoretical physics had reached a kind of standoff with the phenomenon of turbulence. In effect, science had drawn a line on the ground and said, Beyond this we cannot go. On the near side of the line, where fluids behave in orderly ways, there was plenty to work with. Fortunately, a smooth-flowing fluid does not act as though it has a nearly infinite number of independent molecules, each capable of independent motion. Instead, bits of fluid that start nearby tend to remain nearby, like horses in harness. Engineers have workable techniques for calculating flow, as long as it remains calm. They use a body of knowledge dating back to the nineteenth century, when understanding the motions of liquids and gases was a problem on the front lines of physics.

By the modern era, however, it was on the front lines no longer. To the deep theorists, fluid dynamics seemed to retain no mystery but the one that was unapproachable even in heaven. The practical side was so well understood that it could be left to the technicians. Fluid dynamics was no longer really part of physics, the physicists would say. It was mere engineering. Bright young physicists had better things to do. Fluid dynamicists were generally found in university engineering departments. A practical interest in turbulence has always been in the foreground, and the practical interest is usually one-sided: make the turbulence go away. In some applications, turbulence is desirable—inside a jet engine, for example, where efficient burning depends on rapid mixing. But in most, turbulence means disaster. Turbulent airflow over a wing destroys lift. Turbulent flow in an oil pipe creates stupefying drag. Vast amounts of government and corporate money are staked on the design of aircraft, turbine engines, propellers, submarine hulls, and other shapes that move through fluids. Researchers must worry about flow in blood vessels and heart valves. They worry about the shape and evolution of explosions. They worry about vortices and eddies, flames and shock waves. In theory the World War II atomic bomb project was a problem in nuclear physics. In reality the nuclear physics had been mostly solved before the project began, and the business that occupied the scientists assembled at Los Alamos was a problem in fluid dynamics.

What is turbulence then? It is a mess of disorder at all scales, small eddies within large ones. It is unstable. It is highly dissipative, meaning that turbulence drains energy and creates drag. It is motion turned random. But *how* does flow change from smooth to turbulent? Suppose you have a perfectly smooth pipe, with a perfectly even source of water, perfectly shielded from vibrations—how can such a flow create something *random*?

All the rules seem to break down. When flow is smooth, or laminar, small disturbances die out. But past the onset of turbulence, disturbances grow catastrophically. This onset—this transition—became a critical mystery in science. The channel below a rock in a stream becomes a whirling vortex that grows, splits off and spins downstream. A plume of cigarette smoke rises smoothly from an ashtray, accelerating until it passes a critical velocity and splinters into wild eddies. The onset of turbulence can be seen and measured in laboratory experiments; it can be tested for any new wing or propeller by experimental work in a wind tunnel;

but its nature remains elusive. Traditionally, knowledge gained has always been special, not universal. Research by trial and error on the wing of a Boeing 707 aircraft contributes nothing to research by trial and error on the wing of an F-16 fighter. Even supercomputers are close to helpless in the face of irregular fluid motion.

Something shakes a fluid, exciting it. The fluid is viscous—sticky, so that energy drains out of it, and if you stopped shaking, the fluid would naturally come to rest. When you shake it, you add energy at low frequencies, or large wavelengths, and the first thing to notice is that the large wavelengths decompose into small ones. Eddies form, and smaller eddies within them, each dissipating the fluid's energy and each producing a characteristic rhythm. In the 1930s A. N. Kolmogorov put forward a mathematical description that gave some feeling for how these eddies work. He imagined the whole cascade of energy down through smaller and smaller scales until finally a limit is reached, when the eddies become so tiny that the relatively larger effects of viscosity take over.

For the sake of a clean description, Kolmogorov imagined that these eddies fill the whole space of the fluid, making the fluid everywhere the same. This assumption, the assumption of homogeneity, turns out not to be true, and even Poincaré knew it forty years earlier, having seen at the rough surface of a river that the eddies always mix with regions of smooth flow. The vorticity is localized. Energy actually dissipates only in part of the space. At each scale, as you look closer at a turbulent eddy, new regions of calm come into view. Thus the assumption of homogeneity gives way to the assumption of intermittency. The intermittent picture, when idealized somewhat, looks highly fractal, with intermixed regions of roughness and smoothness on scales running down from the large to the small. This picture, too, turns out to fall somewhat short of the reality.

Closely related, but quite distinct, was the question of what happens when turbulence begins. How does a flow cross the boundary from smooth to turbulent? Before turbulence becomes fully developed, what intermediate stages might exist? For these questions, a slightly stronger theory existed. This orthodox paradigm came from Lev D. Landau, the great Russian scientist whose text on fluid dynamics remains a standard. The Landau picture is a piling up of competing rhythms. When more energy comes into a system, he conjectured, new frequencies begin one at a time, each incompatible with the last, as if a violin string responds to harder bowing by vibrating with a second, dissonant tone, and then a third, and a fourth, until the sound becomes an incomprehensible cacophony.

Any liquid or gas is a collection of individual bits, so many that they may as well be infinite. If each piece moved independently, then the fluid would have infinitely many possibilities, infinitely many "degrees of freedom" in the jargon, and the equations describing the motion would have to deal with infinitely many variables. But each particle does not move independently— its motion depends very much on the motion of its neighbors— and in a smooth flow, the degrees of freedom can be few. Potentially complex movements remain coupled together. Nearby bits remain nearby or drift apart in a smooth, linear way that produces neat lines in wind-tunnel pictures. The particles in a column of cigarette smoke rise as one, for a while.

Then confusion appears, a menagerie of mysterious wild motions. Sometimes these motions received names: the oscillatory, the skewed varicose, the cross-roll, the knot, the zigzag. In Landau's view, these unstable new motions simply accumulated, one on top of another, creating rhythms with overlapping speeds and sizes. Conceptually, this orthodox idea of turbulence seemed to fit the facts, and if the theory was mathematically useless—which it was—well, so be it. Landau's paradigm was a way of retaining dignity while throwing up the hands.

Water courses through a pipe, or around a cylinder, making a faint smooth hiss. In your mind, you turn up the pressure. A back-and-forth rhythm begins. Like a wave, it knocks slowly against the pipe. Turn the knob again. From somewhere, a second frequency enters, out of synchronization with the first. The rhythms overlap, compete, jar against one another. Already they create such a complicated motion, waves banging against the walls, interfering with one another, that you almost cannot follow it. Now turn up the knob again. A third frequency enters, then a fourth, a fifth, a sixth, all incommensurate. The flow has become extremely complicated. Perhaps this is turbulence. Physicists accepted this picture, but no one had any idea how to predict when an increase in energy would create a new frequency, or what the new frequency would be. No one had seen these mysteriously arriving frequencies in an experiment because, in fact, no one had ever tested Landau's theory for the onset of turbulence.

THEORISTS CONDUCT EXPERIMENTS with their brains. Experimenters have to use their hands, too. Theorists are thinkers, experimenters are craftsmen. The theorist needs no accomplice. The experimenter has to muster graduate students, cajole machinists, flatter lab assistants. The theorist operates in a pristine place free of noise, of vibration, of dirt. The experimenter develops an intimacy with matter as a sculptor does with clay, battling it, shaping it, and engaging it. The theorist invents his companions, as a naïve Romeo imagined his ideal Juliet. The experimenter's lovers sweat, complain, and fart.

They need each other, but theorists and experimenters have allowed certain inequities to enter their relationships since the ancient days when every scientist was both. Though the best experimenters still have some of the theorist in them, the converse does not hold. Ultimately, prestige accumulates on the theorist's side of the table. In high energy physics, especially, glory goes to the theorists, while experimenters have become highly specialized technicians, managing expensive and complicated equipment. In the decades since World War II, as physics came to be defined by the study of fundamental particles, the best publicized experiments were those carried out with particle accelerators. Spin, symmetry, color, flavor—these were the glamorous abstractions. To most laymen following science, and to more than a few scientists, the study of atomic particles *was* physics. But studying smaller particles, on shorter time scales, meant higher levels of energy. So the machinery needed for good experiments grew with the years, and the nature of experimentation changed for good in particle physics. The field was crowded, and the big experiment encouraged teams. The particle physics papers often stood out in *Physical Review Letters*: a typical authors list could take up nearly one-quarter of a paper's length.

Some experimenters, however, preferred to work alone or in pairs. They worked with substances closer to hand. While such fields as hydrodynamics had lost status, solid-state physics had gained, eventually expanding its territory enough to require a more comprehensive name, "condensed matter physics": the physics of stuff. In condensed matter physics, the machinery was simpler. The gap between theorist and experimenter remained narrower. Theorists expressed a little less snobbery, experimenters a little less defensiveness.

Even so, perspectives differed. It was fully in character for a theorist to interrupt an experimenter's lecture to ask: Wouldn't more data points be more convincing? Isn't that graph a little messy? Shouldn't those numbers extend up and down the scale for a few more orders of magnitude?

And in return, it was fully in character for Harry Swinney to draw himself up to his maximum height, something around five and a half feet, and say, "That's true," with a mixture of innate Louisiana charm and acquired New York irascibility. "That's true if you have an infinite amount of noise-free data." And wheel dismissively back toward the blackboard, adding, "In reality, of course, you have a limited amount of noisy data."

Swinney was experimenting with stuff. For him the turning point had come when he was a graduate student at Johns Hopkins. The excitement of particle physics was palpable. The inspiring Murray Gell-Mann came to talk once, and Swinney was captivated. But when he looked into what graduate students did, he discovered that they were all writing computer programs or soldering spark chambers. It was then that he began talking to an older physicist starting to work on phase transitions—changes from solid to liquid, from nonmagnet to magnet, from conductor to superconductor. Before long Swinney had an empty room—not much bigger than a closet, but it was his alone. He had an equipment catalogue, and he began ordering. Soon he had a table and a laser and some refrigerating equipment and some probes. He designed an apparatus to measure how well carbon dioxide conducted heat around the critical point where it turned from vapor to liquid. Most people thought that the thermal conductivity would change slightly. Swinney found that it changed by a factor of 1,000. That was exciting—alone in a tiny room, discovering something that no one else knew. He saw the other-worldly light that shines from a vapor, any vapor, near the critical point, the light called "opalescence" because the soft scattering of rays gives the white glow of an opal.

Like so much of chaos itself, phase transitions involve a kind of macroscopic behavior that seems hard to predict by looking at the microscopic details. When a solid is heated, its molecules vibrate with the added energy. They push outward against their bonds and force the substance to expand. The more heat, the more expansion. Yet at a certain temperature and pressure, the change becomes sudden and discontinuous. A rope has been stretching; now it breaks. Crystalline form dissolves, and the molecules slide away from one another. They obey fluid laws that could not have been inferred from any aspect of the solid. The average atomic energy has barely changed, but the material—now a liquid, or a magnet, or a superconductor—has entered a new realm.

Günter Ahlers, at AT&T Bell Laboratories in New Jersey, had examined the so-called superfluid transition in liquid helium, in which, as temperature falls, the material becomes a sort of magical flowing liquid with no perceptible viscosity or friction. Others had studied superconductivity. Swinney had studied the critical point where matter changes between liquid and vapor. Swinney, Ahlers, Pierre Bergé, Jerry Gollub, Marzio Giglio—by the middle 1970s these experimenters and others in the United States, France, and Italy, all from the young tradition of exploring phase transitions, were looking for new problems. As intimately as a mail carrier learns the stoops and alleys of his neighborhood, they had learned the peculiar signposts of substances changing their fundamental state. They had studied a brink upon which matter stands poised.

The march of phase transition research had proceeded along stepping stones of analogy: a nonmagnet-magnet phase transition proved to be *like* a liquid-vapor phase transition. The fluid-superfluid phase transition proved to be *like* the conductor-superconductor phase transition. The mathematics of one experiment applied to many other experiments. By the 1970s the problem had been largely solved. A question, though, was how far the theory could be extended. What other changes in the world, when examined closely, would prove to be phase transitions?

It was neither the most original idea nor the most obvious to apply phase transition techniques to flow in fluids. Not the most original because the great hydrodynamic pioneers, Reynolds and Rayleigh and their followers in the early twentieth century, had already noted that a carefully controlled fluid experiment produces a change in the quality of motion—in mathematical terms a bifurcation. In a fluid cell, for example, liquid heated from the bottom suddenly goes from motionlessness to motion. Physicists were tempted to suppose that the physical character of that bifurcation resembled the changes in a substance that fell under the rubric of phase transitions.

It was not the most obvious sort of experiment because, unlike real phase transitions, these fluid bifurcations entailed no change in the substance itself. Instead they added a new element: motion. A still liquid becomes a flowing liquid. Why should the mathematics of such a change correspond to the mathematics of a condensing vapor?

IN 1973 SWINNEY was teaching at the City College of New York. Jerry Gollub, a serious and boyish graduate of Harvard, was teaching at Haverford. Haverford, a mildly bucolic liberal arts college near Philadelphia, seemed less than an ideal place for a physicist to end up. It had no graduate students to help with laboratory work and otherwise fill in the bottom half of the all-important mentor-protégé partnership. Gollub, though, loved teaching undergraduates and began building up the college's physics department into a center widely known for the quality of its experimental work. That year, he took a sabbatical semester and came to New York to collaborate with Swinney.

With the analogy in mind between phase transitions and fluid instabilities, the two men decided to examine a classic system of liquid confined between two vertical cylinders. One cylinder rotated inside the other, pulling the liquid around with it. The system enclosed its flow between surfaces. Thus it restricted the possible motion of the liquid in space, unlike jets and wakes in open water. The rotating cylinders produced what was known as Couette-Taylor flow. Typically, the inner cylinder spins inside a stationary shell, as a matter of convenience. As the rotation begins and picks up speed, the first instability occurs: the liquid forms

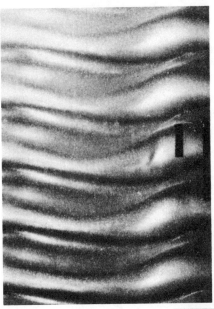

FLOW BETWEEN ROTATING CYLINDERS. The patterned flow of water between two cylinders gave Harry Swinney and Jerry Gollub a way to look at the onset of turbulence. As the rate of spin is increased, the structure grows more complex. First the water forms a characteristic pattern of flow resembling stacked doughnuts. Then the doughnuts begin to ripple. The physicists used a laser to measure the water's changing velocity as each new instability appeared.

an elegant pattern resembling a stack of inner tubes at a service station. Doughnut-shaped bands appear around the cylinder, stacked one atop another. A speck in the fluid rotates not just east to west but also up and in and down and out around the doughnuts. This much was already understood. G. I. Taylor had seen it and measured it in 1923.

To study Couette flow, Swinney and Gollub built an apparatus that fit on a desktop, an outer glass cylinder the size of a skinny can of tennis balls, about a foot high and two inches across. An inner cylinder of steel slid neatly inside, leaving just one-eighth of an inch between for water. "It was a string-and-sealing-wax affair," said Freeman Dyson, one of an unexpected series of prominent sightseers in the months that followed. "You had these two gentlemen in a poky little lab with essentially no money doing an absolutely beautiful experiment. It was the beginning of good quantitative work on turbulence."

The two had in mind a legitimate scientific task that would have brought them a standard bit of recognition for their work and would then have been forgotten. Swinney and Gollub intended to confirm Landau's idea for the onset of turbulence. The experimenters had no reason to doubt it. They knew that fluid dynamicists believed the Landau picture. As physicists they liked it because it fit the general picture of phase transitions, and Landau himself had provided the most workable early framework for studying phase transitions, based on his insight that such phenomena might obey universal laws, with regularities that overrode differences in particular substances. When Harry Swinney studied the liquid-vapor critical point in carbon dioxide, he did so with Landau's conviction that his findings would carry over to the liquid-vapor critical point in xenon—and indeed they did. Why shouldn't turbulence prove to be a steady accumulation of conflicting rhythms in a moving fluid?

Swinney and Gollub prepared to combat the messiness of moving fluids with an arsenal of neat experimental techniques built up over years of studying phase transitions in the most delicate of circumstances. They had laboratory styles and measuring equipment that a fluid dynamicist would never have imagined. To probe the rolling currents, they used laser light. A beam shining through the water would produce a deflection, or scattering, that could be measured in a technique called laser doppler interfero-metry. And the stream of data could be stored and processed by a computer—a device that in 1975 was rarely seen in a tabletop laboratory experiment.

Landau had said new frequencies would appear, one at a time, as a flow increased. "So we read that," Swinney recalled, "and we said, fine, we will look at the transitions where these frequencies come in. So we looked, and sure enough there was a very well-defined transition. We went back and forth through the transition, bringing the rotation speed of the cylinder up and down. It was very well defined."

When they began reporting results, Swinney and Gollub confronted a sociological boundary in science, between the domain of physics and the domain of fluid dynamics. The boundary had certain vivid characteristics. In particular, it determined which bureaucracy within the National Science Foundation controlled their financing. By the 1980s a Couette-Taylor experiment was physics again, but in 1973 it was just plain fluid dynamics, and for people who were accustomed to fluid dynamics, the first numbers coming out of this small City College laboratory were suspiciously clean. Fluid dynamicists just did not believe them. They were not accustomed to experiments in the precise style of phase-transition physics. Furthermore, in the perspective of fluid dynamics, the theoretical point of such an experiment was hard to see. The next time Swinney and Gollub tried to get National Science Foundation money, they were turned down. Some referees did not credit their results, and some said there was nothing new.

But the experiment had never stopped. "There was the transition, very well defined," Swinney said. "So that was great. Then we went on, to look for the next one."

There the expected Landau sequence broke down. Experiment failed to confirm theory. At the next transition the flow jumped all the way to a confused state with no distinguishable cycles at all. No new frequencies, no gradual buildup of complexity. "What we found was, it became chaotic." A few months later, a lean, intensely charming Belgian appeared at the door to their laboratory.

DAVID RUELLE SOMETIMES SAID there were two kinds of physicists, the kind that grew up taking apart radios—this being an era before solid-state, when you could still look at wires and orange-glowing vacuum tubes and imagine something about the flow of electrons—and the kind that played with chemistry sets. Ruelle played with chemistry sets, or not quite sets in the later American sense, but chemicals, explosive or poisonous, cheerfully dispensed in his native northern Belgium by the local pharmacist and then mixed, stirred, heated, crystallized, and sometimes blown up by Ruelle himself. He was born in Ghent in 1935, the son of a gymnastics teacher and a university professor of linguistics, and though he made his career in an abstract realm of science he always had a taste for a dangerous side of nature that hid its surprises in cryptogamous fungoid mushrooms or saltpeter, sulfur, and charcoal.

It was in mathematical physics, though, that Ruelle made his lasting contribution to the exploration of chaos. By 1970 he had joined the Institut des Hautes Études Scientifiques, an institute outside Paris modeled on the Institute for Advanced Study in Princeton. He had already developed what became a lifelong habit of leaving the institute and his family periodically to take solitary walks, weeks long, carrying only a backpack through empty wildernesses in Iceland or rural Mexico. Often he saw no one. When he came across humans and accepted their hospitality—perhaps a meal of maize tortillas, with no fat, animal or vegetable—he felt that he was seeing the world as it existed two millennia before. When he returned to the institute he would begin his scientific existence again, his face just a little more gaunt, the skin stretched a little more tightly over his round brow and sharp chin. Ruelle had heard talks by Steve Smale about the horseshoe map and the chaotic possibilities of dynamical systems. He had also thought about fluid turbulence and the classic Landau picture. He suspected that these ideas were related—and contradictory.

Ruelle had no experience with fluid flows, but that did not discourage him any more than it had discouraged his many unsuccessful predecessors. "Always nonspecialists find the new things," he said. "There is not a natural deep theory of turbulence. All the questions you can ask about turbulence are of a more general nature, and therefore accessible to nonspecialists." It was easy to see why turbulence resisted analysis. The equations of fluid flow are nonlinear partial differential equations, unsolvable except in special cases. Yet Ruelle worked out an abstract alternative to Landau's picture, couched in the language of Smale, with images of space as a pliable material to be squeezed, stretched, and folded into shapes like horseshoes. He wrote a paper at his institute with a visiting Dutch mathematician, Floris Takens, and they published it in 1971. The style was unmistakably mathematics—physicists, beware!—meaning that paragraphs would begin with *Definition* or *Proposition* or *Proof*, followed by the inevitable thrust: *Let* . . . .

"**Proposition (5.2).** *Let* $X\mu$ *be a one-parameter family of* C^k *vectorfields on a Hilbert space H such that* . . ."

Yet the title claimed a connection with the real world: "On the Nature of Turbulence," a deliberate echo of Landau's famous title, "On the Problem of Turbulence." The clear purpose of Ruelle and Takens's argument went beyond mathematics; they meant to offer a substitute for the traditional view of the onset of turbulence. Instead of a piling up of frequencies, leading to an infinitude of independent overlapping motions, they proposed that just three independent motions would produce the full complexity of turbulence. Mathematically speaking, some of their logic turned out to be obscure, wrong, borrowed, or all three—opinions still varied fifteen years later.

But the insight, the commentary, the marginalia, and the physics woven into the paper made it a lasting gift. Most seductive of all was an image that the authors called a *strange attractor*. This phrase was psychoanalytically "suggestive," Ruelle felt later. Its status in the study of chaos was such that he and Takens jousted below a polite surface for the honor of having chosen the words. The truth was that neither quite remembered, but Takens, a tall, ruddy, fiercely Nordic man, might say, "Did you ever ask God whether he created this damned universe? . . . I don't remember anything. . . . I often create without remembering it," while Ruelle, the paper's senior author, would remark softly, "Takens happened to be visiting IHES. Different people work differently. Some people would try to write a paper all by themselves so they keep all the credit."

The strange attractor lives in phase space, one of the most powerful inventions of modern science. Phase space gives a way of turning numbers into pictures, abstracting every bit of essential information from a system of moving parts, mechanical or fluid, and making a flexible road map to all its possibilities. Physicists already worked with two simpler kinds of "attractors": fixed points and limit cycles, representing behavior that reached a steady state or repeated itself continuously.

In phase space the complete state of knowledge about a dynamical system at a single instant in time collapses to a point. That point *is* the dynamical system—at that instant. At the next instant, though, the system will have changed, ever so slightly, and so the point moves. The history of the system time can be charted by the moving point, tracing its orbit through phase space with the passage of time.

How can all the information about a complicated system be stored in a point? If the system has only two variables, the answer is simple. It is straight from the Cartesian geometry taught in high school—one variable on the horizontal axis, the other on the vertical. If the system is a swinging, frictionless pendulum, one variable is position and the other velocity, and they change continuously, making a line of points that traces a loop, repeating itself forever, around and around. The same system with a higher energy level—swinging faster and farther—forms a loop in phase space similar to the first, but larger.

A little realism, in the form of friction, changes the picture. We do not need the equations of motion to know the destiny of a pendulum subject to friction. Every orbit must eventually end up at the same place, the center: position 0, velocity 0. This central fixed point "attracts" the orbits. Instead of looping around forever, they spiral inward. The friction dissipates the system's energy, and in phase space the dissipation shows itself as a pull toward the center, from the outer regions of high energy to the inner regions of low energy. The attractor—the simplest kind possible—is like a pinpoint magnet embedded in a rubber sheet.

One advantage of thinking of states as points in space is that it makes change easier to watch. A system whose variables change continuously up or down becomes a moving point, like a fly moving around a room. If some combinations of variables never occur, then a scientist can simply imagine that part of the room as out of bounds. The fly never goes there. If a system behaves periodically, coming around to the same state again and again, then the fly moves in a loop, passing through the same position in phase space again and again. Phase-space portraits of physical systems exposed patterns of motion that were invisible otherwise, as an infrared landscape photograph can reveal patterns and details that exist just beyond the reach of perception. When a scientist looked at a phase portrait, he could use his imagination to think back to the system itself. This loop corresponds to that periodicity. This twist corresponds to that change. This empty void corresponds to that physical impossibility.

Even in two dimensions, phase-space portraits had many surprises in store, and even desktop computers could easily demonstrate some of them, turning equations into colorful moving trajectories. Some physicists began making movies and videotapes to show their colleagues, and some mathematicians in California published books with a series of green, blue, and red cartoon-style drawings—"chaos comics," some of their colleagues said, with just a touch of malice. Two dimensions did not begin to cover the kinds of systems that physicists needed to study. They had to show more variables than two, and that meant more dimensions. Every piece of a dynamical system that can move independently is another variable, another degree of freedom. Every degree of freedom requires another dimension in phase space, to make sure that a single point contains enough information to determine the state of the system uniquely. The simple equations Robert May studied were one-dimensional—a single number was enough, a number that might stand for temperature or population, and that number defined the position of a point on a one-dimensional line. Lorenz's stripped-down system of fluid convection was three-dimensional, not because the fluid moved through three dimensions, but because it took three distinct numbers to nail down the state of the fluid at any instant.

Spaces of four, five, or more dimensions tax the visual imagination of even the most agile topologist. But complex systems

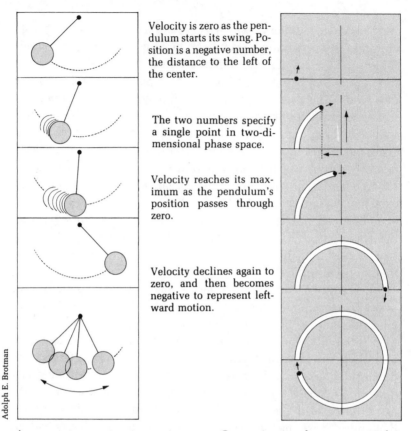

Velocity is zero as the pendulum starts its swing. Position is a negative number, the distance to the left of the center.

The two numbers specify a single point in two-dimensional phase space.

Velocity reaches its maximum as the pendulum's position passes through zero.

Velocity declines again to zero, and then becomes negative to represent leftward motion.

Adolph E. Brotman

ANOTHER WAY TO SEE A PENDULUM. One point in phase space (right) contains all the information about the state of a dynamical system at any instant (left). For a simple pendulum, two numbers—velocity and position—are all you need to know.

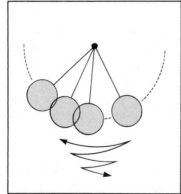

Adolph E. Brotman

The points trace a trajectory that provides a way of visualizing the continuous long-term behavior of a dynamical system. A repeating loop represents a system that repeats itself at regular intervals forever.

If the repeating behavior is stable, as in a pendulum clock, then the system returns to this orbit after small perturbations. In phase space, trajectories near the orbit are drawn into it; the orbit is an attractor.

have many independent variables. Mathematicians had to accept the fact that systems with infinitely many degrees of freedom— untrammeled nature expressing itself in a turbulent waterfall or an unpredictable brain—required a phase space of infinite dimensions. But who could handle such a thing? It was a hydra, merciless and uncontrollable, and it was Landau's image for turbulence: infinite modes, infinite degrees of freedom, infinite dimensions.

A PHYSICIST HAD GOOD REASON to dislike a model that found so little clarity in nature. Using the nonlinear equations of fluid motion, the world's fastest supercomputers were incapable of accurately tracking a turbulent flow of even a cubic centimeter for more than a few seconds. The blame for this was certainly nature's more than Landau's, but even so the Landau picture went against the grain. Absent any knowledge, a physicist might be permitted to suspect that some principle was evading discovery. The great quantum theorist Richard P. Feynman expressed this feeling. "It always bothers me that, according to the laws as we understand them today, it takes a computing machine an infinite number of logical operations to figure out what goes on in no matter how tiny a region of space, and no matter how tiny a region of time. How can all that be going on in that tiny space? Why should it take an infinite amount of logic to figure out what one tiny piece of space/time is going to do?"

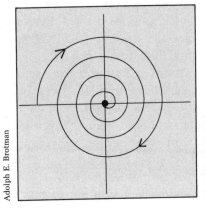

Adolph E. Brotman

An attractor can be a single point. For a pendulum steadily losing energy to friction, all trajectories spiral inward toward a point that represents a steady state—in this case, the steady state of no motion at all.

Like so many of those who began studying chaos, David Ruelle suspected that the visible patterns in turbulent flow—self-entangled stream lines, spiral vortices, whorls that rise before the eye and vanish again—must reflect patterns explained by laws not yet discovered. In his mind, the dissipation of energy in a turbulent flow must still lead to a kind of contraction of the phase space, a pull toward an attractor. Certainly the attractor would not be a fixed point, because the flow would never come to rest. Energy was pouring into the system as well as draining out. What other kind of attractor could it be? According to dogma, only one other kind existed, a periodic attractor, or limit cycle—an orbit that attracted all other nearby orbits. If a pendulum gains energy from a spring while it loses it through friction—that is, if the pendulum is driven as well as damped—a stable orbit may be the closed loop in phase space that represents the regular swinging motion of a grandfather clock. No matter where the pendulum starts, it will settle into that one orbit. Or will it? For some initial conditions—those with the lowest energy—the pendulum will still settle to a stop, so the system actually has two attractors, one a closed loop and the other a fixed point. Each attractor has its "basin," just as two nearby rivers have their own watershed regions.

In the short term any point in phase space can stand for a possible behavior of the dynamical system. In the long term the only possible behaviors are the attractors themselves. Other kinds of motion are transient. By definition, attractors had the important property of stability—in a real system, where moving parts are subject to bumps and jiggles from real-world noise, motion tends to return to the attractor. A bump may shove a trajectory away for a brief time, but the resulting transient motions die out. Even if the cat knocks into it, a pendulum clock does not switch to a sixty-two-second minute. Turbulence in a fluid was a behavior of a different order, never producing any single rhythm to the exclusion of others. A well-known characteristic of turbulence was that the whole broad spectrum of possible cycles was present at once. Turbulence is like white noise, or static. Could such a thing arise from a simple, deterministic system of equations?

Ruelle and Takens wondered whether some other kind of attractor could have the right set of properties. Stable—representing the final state of a dynamical system in a noisy world.

Low-dimensional—an orbit in a phase space that might be a rectangle or a box, with just a few degrees of freedom. Nonperiodic— never repeating itself, and never falling into a steady grandfather-clock rhythm. Geometrically the question was a puzzle: What kind of orbit could be drawn in a limited space so that it would never repeat itself and never cross itself—because once a system returns to a state it has been in before, it thereafter must follow the same path. To produce *every* rhythm, the orbit would have to be an infinitely long line in a finite area. In other words—but the word had not been invented—it would have to be fractal.

By mathematical reasoning, Ruelle and Takens claimed that such a thing must exist. They had never seen one, and they did not draw one. But the claim was enough. Later, delivering a plenary address to the International Congress of Mathematicians in Warsaw, with the comfortable advantage of hindsight, Ruelle declared: "The reaction of the scientific public to our proposal was quite cold. In particular, the notion that continuous spectrum would be associated with a few degrees of freedom was viewed as heretical by many physicists." But it was physicists—a handful, to be sure—who recognized the importance of the 1971 paper and went to work on its implications.

ACTUALLY, BY 1971 the scientific literature already contained one small line drawing of the unimaginable beast that Ruelle and Takens were trying to bring alive. Edward Lorenz had attached it to his 1963 paper on deterministic chaos, a picture with just two curves on the right, one inside the other, and five on the left. To plot just these seven loops required 500 successive calculations on the computer. A point moving along this trajectory in phase space, around the loops, illustrated the slow, chaotic rotation of a fluid as modeled by Lorenz's three equations for convection. Because the system had three independent variables, this attractor lay in a three-dimensional phase space. Although Lorenz drew only a fragment of it, he could see more than he drew: a sort of double spiral, like a pair of butterfly wings interwoven with infinite dexterity. When the rising heat of his system pushed the fluid around in one direction, the trajectory stayed on the right wing; when the rolling motion stopped and reversed itself, the trajectory would swing across to the other wing.

The attractor was stable, low-dimensional, and nonperiodic. It could never intersect itself, because if it did, returning to a point already visited, from then on the motion would repeat itself in a periodic loop. That never happened—that was the beauty of the attractor. Those loops and spirals were infinitely deep, never quite joining, never intersecting. Yet they stayed inside a finite space, confined by a box. How could that be? How could infinitely many paths lie in a finite space?

In an era before Mandelbrot's pictures of fractals had flooded the scientific marketplace, the details of constructing such a shape were hard to imagine, and Lorenz acknowledged an "apparent contradiction" in his tentative description. "It is difficult to reconcile the merging of two surfaces, one containing each spiral, with the inability of two trajectories to merge," he wrote. But he saw an answer too delicate to appear in the few calculations within range of his computer. Where the spirals appear to join, the sur-

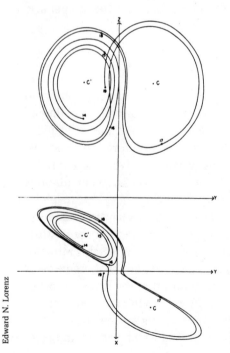

Edward N. Lorenz

THE FIRST STRANGE ATTRACTOR. In 1963 Edward Lorenz could compute only the first few strands of the attractor for his simple system of equations. But he could see that the interleaving of the two spiral wings must have an extraordinary structure on invisibly small scales.

faces must divide, he realized, forming separate layers in the manner of a flaky mille-feuille. "We see that each surface is really a pair of surfaces, so that, where they appear to merge, there are really four surfaces. Continuing this process for another circuit, we see that there are really eight surfaces, etc., and we finally conclude that there is an infinite complex of surfaces, each extremely close to one or the other of two merging surfaces." It was no wonder that meteorologists in 1963 left such speculation alone, nor that Ruelle a decade later felt astonishment and excitement when he finally learned of Lorenz's work. He went to visit Lorenz once, in the years that followed, and left with a small sense of disappointment that they had not talked more of their common territory in science. With characteristic diffidence, Lorenz made the occasion a social one, and they went with their wives to an art museum.

The effort to pursue the hints put forward by Ruelle and Takens took two paths. One was the theoretical struggle to visualize strange attractors. Was the Lorenz attractor typical? What other sorts of shapes were possible? The other was a line of experimental work meant to confirm or refute the highly unmathematical leap of faith that suggested the applicability of strange attractors to chaos in nature.

In Japan the study of electrical circuits that imitated the behavior of mechanical springs—but much faster—led Yoshisuke Ueda to discover an extraordinarily beautiful set of strange attractors. (He met an Eastern version of the coolness that greeted Ruelle: "Your result is no more than an almost periodic oscillation. Don't form a selfish concept of steady states.") In Germany Otto Rössler, a nonpracticing medical doctor who came to chaos by way of chemistry and theoretical biology, began with an odd ability to see strange attractors as philosophical objects, letting the mathematics follow along behind. Rössler's name became attached to a particularly simple attractor in the shape of a band of ribbon with a fold in it, much studied because it was easy to draw, but he also visualized attractors in higher dimensions—"a sausage in a sausage in a sausage in a sausage," he would say, "take it out, fold it, squeeze it, put it back." Indeed, the folding and squeezing of space was a key to constructing strange attractors, and perhaps a key to the dynamics of the real systems that gave rise to them.

Rössler felt that these shapes embodied a self-organizing principle in the world. He would imagine something like a wind sock on an airfield, "an open hose with a hole in the end, and the wind forces its way in," he said. "Then the wind is trapped. Against its will, energy is doing something productive, like the devil in medieval history. The principle is that nature does something against its own will and, by self-entanglement, produces beauty."

Making pictures of strange attractors was not a trivial matter. Typically, orbits would wind their ever-more-complicated paths through three dimensions or more, creating a dark scribble in space with an internal structure that could not be seen from the outside. To convert these three-dimensional skeins into flat pictures, scientists first used the technique of projection, in which a drawing represented the shadow that an attractor would cast on a surface. But with complicated strange attractors, projection just smears the detail into an indecipherable mess. A more revelatory technique was to make a *return map*, or a *Poincaré map*, in effect, taking a slice from the tangled heart of the attractor, removing a two-dimensional section just as a pathologist prepares a section of tissue for a microscope slide.

The Poincaré map removes a dimension from an attractor and turns a continuous line into a collection of points. In reducing an attractor to its Poincaré map, a scientist implicitly assumes that he can preserve much of the essential movement. He can imagine, for example, a strange attractor buzzing around before his eyes, its orbits carrying up and down, left and right, and to and fro through his computer screen. Each time the orbit passes through the screen, it leaves a glowing point at the place of intersection, and the points either form a random blotch or begin to trace some shape in phosphorus.

The process corresponds to sampling the state of a system every so often, instead of continuously. When to sample—where to take the slice from a strange attractor—is a question that gives an investigator some flexibility. The most informative interval might correspond to some physical feature of the dynamical system: for example, a Poincaré map could sample the velocity of a pendulum bob each time it passed through its lowest point. Or the investigator could choose some regular time interval, freezing

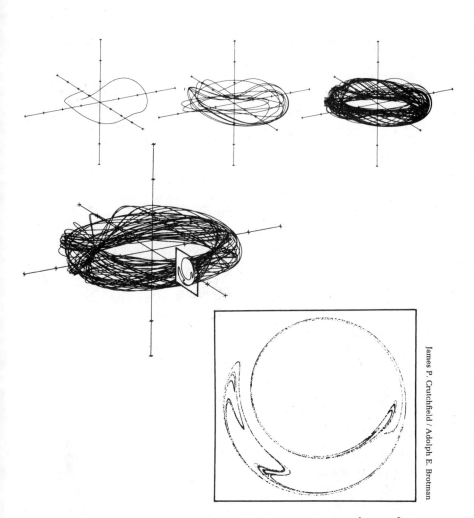

EXPOSING AN ATTRACTOR'S STRUCTURE. The strange attractor above—first one orbit, then ten, then one hundred—depicts the chaotic behavior of a rotor, a pendulum swinging through a full circle, driven by an energetic kick at regular intervals. By the time 1,000 orbits have been drawn (*below*), the attractor has become an impenetrably tangled skein.

To see the structure within, a computer can take a slice through an attractor, a so-called Poincaré section. The technique reduces a three-dimensional picture to two dimensions. Each time the trajectory passes through a plane, it marks a point, and gradually a minutely detailed pattern emerges. This example has more than 8,000 points, each standing for a full orbit around the attractor. In effect, the system is "sampled" at regular intervals. One kind of information is lost; another is brought out in high relief.

successive states in the flash of an imaginary strobe light. Either way, such pictures finally began to reveal the fine fractal structure guessed at by Edward Lorenz.

THE MOST ILLUMINATING STRANGE ATTRACTOR, because it was the simplest, came from a man far removed from the mysteries of turbulence and fluid dynamics. He was an astronomer, Michel Hénon of the Nice Observatory on the southern coast of France. In one way, of course, astronomy gave dynamical systems its start, the clockwork motions of planets providing Newton with his triumph and Laplace with his inspiration. But celestial mechanics differed from most earthly systems in a crucial respect. Systems that lose energy to friction are dissipative. Astronomical systems are not: they are conservative, or Hamiltonian. Actually, on a nearly infinitesimal scale, even astronomical systems suffer a kind of drag, with stars radiating away energy and tidal friction draining some momentum from orbiting bodies, but for practical purposes, astronomers' calculations could ignore dissipation. And without dissipation, the phase space would not fold and contract in the way needed to produce an infinite fractal layering. A strange attractor could never arise. Could chaos?

Many astronomers have long and happy careers without giving dynamical systems a thought, but Hénon was different. He was born in Paris in 1931, a few years younger than Lorenz but, like him, a scientist with a certain unfulfilled attraction to mathematics. Hénon liked small, concrete problems that could be attached to physical situations—"not like the kind of mathematics people do today," he would say. When computers reached a size suitable for hobbyists, Hénon got one, a Heathkit that he soldered together and played with at home. Long before that, though, he took on a particularly baffling problem in dynamics. It concerned globular clusters—crowded balls of stars, sometimes a million in one place, that form the oldest and possibly the most breathtaking objects in the night sky. Globular clusters are amazingly dense with stars. The problem of how they stay together and how they evolve over time has perplexed astronomers throughout the twentieth century.

Dynamically speaking, a globular cluster is a big many-body

problem. The two-body problem is easy. Newton solved it completely. Each body—the earth and the moon, for example—travels in a perfect ellipse around the system's joint center of gravity. Add just one more gravitational object, however, and everything changes. The three-body problem is hard, and worse than hard. As Poincaré discovered, it is most often impossible. The orbits can be calculated numerically for a while, and with powerful computers they can be tracked for a long while before uncertainties begin to take over. But the equations cannot be solved analytically, which means that long-term questions about a three-body system cannot be answered. Is the solar system stable? It certainly appears to be, in the short term, but even today no one knows for sure that some planetary orbits could not become more and more eccentric until the planets fly off from the system forever.

A system like a globular cluster is far too complex to be treated directly as a many-body problem, but its dynamics can be studied with the help of certain compromises. It is reasonable, for example, to think of individual stars winging their way through an average gravitational field with a particular gravitational center. Every so often, however, two stars will approach each other closely enough that their interaction must be treated separately. And astronomers realized that globular clusters generally must not be stable. Binary star systems tend to form inside them, stars pairing off in tight little orbits, and when a third star encounters a binary, one of the three tends to get a sharp kick. Every so often, a star will gain enough energy from such an interaction to reach escape velocity and depart the cluster forever; the rest of the cluster will then contract slightly. When Hénon took on this problem for his doctoral thesis in Paris in 1960, he made a rather arbitrary assumption: that as the cluster changed scale, it would remain self-similar. Working out the calculations, he reached an astonishing result. The core of a cluster would collapse, gaining kinetic energy and seeking a state of infinite density. This was hard to imagine, and furthermore it was not supported by the evidence of clusters so far observed. But slowly Hénon's theory—later given the name "gravothermal collapse"—took hold.

Thus fortified, willing to try mathematics on old problems and willing to pursue unexpected results to their unlikely outcomes, he began work on a much easier problem in star dynamics.

This time, in 1962, visiting Princeton University, he had access for the first time to computers, just as Lorenz at M.I.T. was starting to use computers in meteorology. Hénon began modeling the orbits of stars around their galactic center. In reasonably simple form, galactic orbits can be treated like the orbits of planets around a sun, with one exception: the central gravity source is not a point, but a disk with thickness in three dimensions.

He made a compromise with the differential equations. "To have more freedom of experimentation," as he put it, "we forget momentarily about the astronomical origin of the problem." Although he did not say so at the time, "freedom of experimentation" meant, in part, freedom to play with the problem on a primitive computer. His machine had less than a thousandth of the memory on a single chip of a personal computer twenty-five years later, and it was slow, too. But like later experimenters in the phenomena of chaos, Hénon found that the oversimplification paid off. By abstracting only the essence of his system, he made discoveries that applied to other systems as well, and more important systems. Years later, galactic orbits were still a theoretical game, but the dynamics of such systems were under intense, expensive investigation by those interested in the orbits of particles in high-energy accelerators and those interested in the confinement of magnetic plasmas for the creation of nuclear fusion.

Stellar orbits in galaxies, on a time scale of some 200 million years, take on a three-dimensional character instead of making perfect ellipses. Three-dimensional orbits are as hard to visualize when the orbits are real as when they are imaginary constructions in phase space. So Hénon used a technique comparable to the making of Poincaré maps. He imagined a flat sheet placed upright on one side of the galaxy so that every orbit would sweep through it, as horses on a race track sweep across the finish line. Then he would mark the point where the orbit crossed this plane and trace the movement of the point from orbit to orbit.

Hénon had to plot these points by hand, but eventually the many scientists using this technique would watch them appear on a computer screen, like distant street lamps coming on one by one at nightfall. A typical orbit might begin with a point toward the lower left of the page. Then, on the next go-round, a point would appear a few inches to the right. Then another, more to the right and up a little—and so on. At first no pattern would be obvious, but after ten or twenty points an egg-shaped curve would take shape. The successive points actually make a circuit around the curve, but since they do not come around to exactly the same place, eventually, after hundreds or thousands of points, the curve is solidly outlined.

Such orbits are not completely regular, since they never exactly repeat themselves, but they are certainly predictable, and they are far from chaotic. Points never arrive inside the curve or outside it. Translated back to the full three-dimensional picture, the orbits were outlining a torus, or doughnut shape, and Hénon's mapping was a cross-section of the torus. So far, he was merely illustrating what all his predecessors had taken for granted. Orbits were periodic. At the observatory in Copenhagen, from 1910 to 1930, a generation of astronomers painstakingly observed and calculated hundreds of such orbits—but they were only interested in the ones that proved periodic. "I, too, was convinced, like everyone else at that time, that all orbits should be regular like this," Hénon said. But he and his graduate student at Princeton, Carl Heiles, continued computing different orbits, steadily increasing the level of energy in their abstract system. Soon they saw something utterly new.

First the egg-shaped curve twisted into something more complicated, crossing itself in figure eights and splitting apart into separate loops. Still, every orbit fell on some loop. Then, at even higher levels, another change occurred, quite abruptly. "Here comes the surprise," Hénon and Heiles wrote. Some orbits became so unstable that the points would scatter randomly across the paper. In some places, curves could still be drawn; in others, no curve fit the points. The picture became quite dramatic: evidence of complete disorder mixed with the clear remnants of order, forming shapes that suggested "islands" and "chains of islands" to these astronomers. They tried two different computers and two different methods of integration, but the results were the same. They could only explore and speculate. Based solely on their numerical experimentation, they made a guess about the deep structure of such pictures. With greater magnification, they suggested, more islands

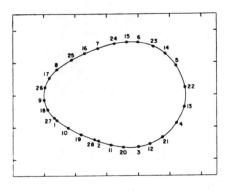

ORBITS AROUND THE GALACTIC CENTER. To understand the trajectories of the stars through a galaxy, Michel Hénon computed the intersections of an orbit with a plane. The resulting patterns depended on the system's total energy. The points from a stable orbit gradually produced a continuous, connected curve (*left*). Other energy levels, however, produced complicated mixtures of stability and chaos, represented by regions of scattered points.

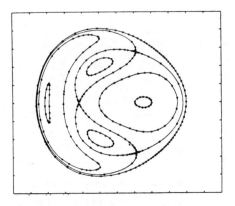

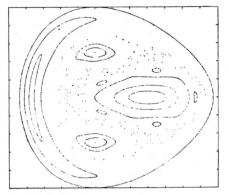

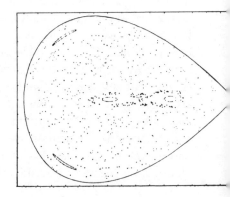

Michel Hénon

would appear on smaller and smaller scales, perhaps all the way to infinity. Mathematical proof was needed—"but the mathematical approach to the problem does not seem too easy."

Hénon went on to other problems, but fourteen years later, when finally he heard about the strange attractors of David Ruelle and Edward Lorenz, he was prepared to listen. By 1976 he had moved to the Observatory of Nice, perched high above the Mediterranean Sea on the Grande Corniche, and he heard a talk by a visiting physicist about the Lorenz attractor. The physicist had been trying different techniques to illuminate the fine "microstructure" of the attractor, with little success. Hénon, though dissipative systems were not his field ("sometimes astronomers are fearful of dissipative systems—they're untidy"), thought he had an idea.

Once again, he decided to throw out all reference to the physical origins of the system and concentrate only on the geometrical essence he wanted to explore. Where Lorenz and others had stuck to differential equations—flows, with *continuous* changes in space and time—he turned to difference equations, discrete in time. The key, he believed, was the repeated stretching and folding of phase space in the manner of a pastry chef who rolls the dough, folds it, rolls it out again, folds it, creating a structure that will eventually be a sheaf of thin layers. Hénon drew a flat oval on a piece of paper. To stretch it, he picked a short numerical function that would move any point in the oval to a new point in a shape that was stretched upward in the center, an arch. This was a mapping—point by point, the entire oval was "mapped" onto the arch. Then he chose a second mapping, this time a contraction that would shrink the arch inward to make it narrower. And then a third mapping turned the narrow arch on its side, so that it would line up neatly with the original oval. The three mappings could be combined into a single function for purposes of calculation.

In spirit he was following Smale's horseshoe idea. Numerically, the whole process was so simple that it could easily be tracked on a calculator. Any point has an x coordinate and a y coordinate to fix its horizontal and vertical position. To find the new x, the rule was to take the old y, add 1 and subtract 1.4 times the old x squared. To find the new y, multiply 0.3 by the old x. That is: $x_{new} = y + 1 - 1.4x^2$ and $y_{new} = 0.3x$. Hénon picked a starting point more or less at random, took his calculator and started plotting new points, one after another, until he had plotted thousands. Then he used a real computer, an IBM 7040, and quickly plotted five million. Anyone with a personal computer and a graphics display could easily do the same.

At first the points appear to jump randomly around the screen. The effect is that of a Poincaré section of a three-dimensional attractor, weaving erratically back and forth across the display. But quickly a shape begins to emerge, an outline curved like a banana. The longer the program runs, the more detail appears. Parts of the outline seem to have some thickness, but then the thickness resolves itself into two distinct lines, then the two into four, one pair close together and one pair farther apart. On greater magnification, each of the four lines turns out to be composed of two more lines—and so on, ad infinitum. Like Lorenz's attractor, Hénon's displays infinite regress, like an unending sequence of Russian dolls one inside the other.

The nested detail, lines within lines, can be seen in final form in a series of pictures with progressively greater magnification. But the eerie effect of the strange attractor can be appreciated another way when the shape emerges in time, point by point. It appears like a ghost out of the mist. New points scatter so randomly across the screen that it seems incredible that any structure is there, let alone a structure so intricate and fine. Any two consecutive points are arbitrarily far apart, just like any two points initially nearby in a turbulent flow. Given any number of points, it is impossible to guess where the next will appear—except, of course, that it will be somewhere on the attractor.

The points wander so randomly, the pattern appears so ethereally, that it is hard to remember that the shape is an *attractor*. It is not just any trajectory of a dynamical system. It is the trajectory toward which all other trajectories converge. That is why the choice of starting conditions does not matter. As long as the starting point lies somewhere near the attractor, the next few points will converge to the attractor with great rapidity.

YEARS BEFORE, WHEN DAVID RUELLE arrived at the City College laboratory of Gollub and Swinney in 1974, the three physicists

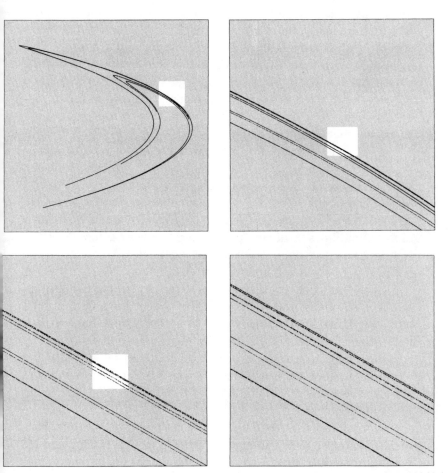

THE ATTRACTOR OF HÉNON. A simple combination of folding and stretching produced an attractor that easy to compute yet still poorly understood by mathematicians. As thousands, the millions of points appear, more and more detail emerges. What appear to be single lines prove, on magnification, to be pairs, then pairs of pairs. Yet whether any two successive points appear nearby or far apart is unpredictable.

found themselves with a slender link between theory and experiment. One piece of mathematics, philosophically bold but technically uncertain. One cylinder of turbulent fluid, not much to look at, but clearly out of harmony with the old theory. The men spent the afternoon talking, and then Swinney and Gollub left for a vacation with their wives in Gollub's cabin in the Adirondack mountains. They had not seen a strange attractor, and they had not measured much of what might actually happen at the onset of turbulence. But they knew that Landau was wrong, and they suspected that Ruelle was right.

As an element in the world revealed by computer exploration, the strange attractor began as a mere possibility, marking a place where many great imaginations in the twentieth century had failed to go. Soon, when scientists saw what computers had to show, it seemed like a face they had been seeing everywhere, in the music of turbulent flows or in clouds scattered like veils across the sky. Nature was *constrained*. Disorder was channeled, it seemed, into patterns with some common underlying theme.

Later, the recognition of strange attractors fed the revolution in chaos by giving numerical explorers a clear program to carry out. They looked for strange attractors everywhere, wherever nature seemed to be behaving randomly. Many argued that the earth's weather might lie on a strange attractor. Others assembled millions of pieces of stock market data and began searching for a strange attractor there, peering at randomness through the adjustable lens of a computer.

In the middle 1970s these discoveries lay in the future. No one had actually seen a strange attractor in an experiment, and it was far from clear how to go about looking for one. In theory the strange attractor could give mathematical substance to fundamental new properties of chaos. Sensitive dependence on initial conditions was one. "Mixing" was another, in a sense that would be meaningful to a jet engine designer, for example, concerned about the efficient combination of fuel and oxygen. But no one knew how to measure these properties, how to attach numbers to them. Strange attractors seemed fractal, implying that their true dimension was fractional, but no one knew how to measure the dimension or how to apply such a measurement in the context of engineering problems.

Most important, no one knew whether strange attractors would say anything about the deepest problem with nonlinear systems. Unlike linear systems, easily calculated and easily classified, nonlinear systems still seemed, in their essence, beyond classification—each different from every other. Scientists might begin to suspect that they shared common properties, but when it came time to make measurements and perform calculations, each nonlinear system was a world unto itself. Understanding one seemed to offer no help in understanding the next. An attractor like Lorenz's illustrated the stability and the hidden structure of a system that otherwise seemed patternless, but how did this peculiar double spiral help researchers exploring unrelated systems? No one knew.

For now, the excitement went beyond pure science. Scientists who saw these shapes allowed themselves to forget momentarily the rules of scientific discourse. Ruelle, for example: "I have not spoken of the esthetic appeal of strange attractors. These systems of curves, these clouds of points suggest sometimes fireworks or galaxies, sometimes strange and disquieting vegetal proliferations. A realm lies there of forms to explore, and harmonies to discover."

Universality

The iterating of these lines brings gold;
The framing of this circle on the ground
Brings whirlwinds, tempests, thunder and lightning.

—MARLOWE, Dr. Faustus

A FEW DOZEN YARDS upstream from a waterfall, a smooth flowing stream seems to intuit the coming drop. The water begins to speed and shudder. Individual rivulets stand out like coarse, throbbing veins. Mitchell Feigenbaum stands at streamside. He is sweating slightly in sports coat and corduroys and puffing on a cigarette. He has been walking with friends, but they have gone on ahead to the quieter pools upstream. Suddenly, in what might be a demented high-speed parody of a tennis spectator, he starts turning his head from side to side. "You can focus on something, a bit of foam or something. If you move your head fast enough, you can all of a sudden discern the whole structure of the surface, and you can feel it in your stomach." He draws in more smoke from his cigarette. "But for anyone with a mathematical background, if you look at this stuff, or you see clouds with all their puffs on top of puffs, or you stand at a sea wall in a storm, you know that you really don't know anything."

Order in chaos. It was science's oldest cliché. The idea of hidden unity and common underlying form in nature had an intrinsic appeal, and it had an unfortunate history of inspiring pseudoscientists and cranks. When Feigenbaum came to Los Alamos National Laboratory in 1974, a year shy of his thirtieth birthday, he knew that if physicists were to make something of the idea now, they would need a practical framework, a way to turn ideas into calculations. It was far from obvious how to make a first approach to the problem.

Feigenbaum was hired by Peter Carruthers, a calm, deceptively genial physicist who came from Cornell in 1973 to take over
the Theoretical Division. His first act was to dismiss a half-dozen
senior scientists—Los Alamos provides its staff with no equivalent
of university tenure—and to replace them with some bright young
researchers of his own choosing. As a scientific manager, he had
strong ambition, but he knew from experience that good science
cannot always be planned.

"If you had set up a committee in the laboratory or in Washington and said, 'Turbulence is really in our way, we've got to
understand it, the lack of understanding really destroys our chance
of making progress in a lot of fields,' then, of course, you would
hire a team. You'd get a giant computer. You'd start running big
programs. And you would never get anywhere. Instead we have
this smart guy, sitting quietly—talking to people, to be sure, but
mostly working all by himself." They had talked about turbulence,
but time passed, and even Carruthers was no longer sure where
Feigenbaum was headed. "I thought he had quit and found a
different problem. Little did I know that this other problem was
the *same* problem. It seems to have been the issue on which many
different fields of science were stuck—they were stuck on this
aspect of the nonlinear behavior of systems. Now, nobody would
have thought that the right background for this problem was to
know particle physics, to know something about quantum field
theory, and to know that in quantum field theory you have these
structures known as the renormalization group. Nobody knew that
you would need to understand the general theory of stochastic
processes, and also fractal structures.

"Mitchell had the right background. He did the right thing at
the right time, and he did it very well. Nothing partial. He cleaned
out the whole problem."

Feigenbaum brought to Los Alamos a conviction that his science had failed to understand hard problems—nonlinear problems. Although he had produced almost nothing as a physicist,
he had accumulated an unusual intellectual background. He had
a sharp working knowledge of the most challenging mathematical
analysis, new kinds of computational technique that pushed most
scientists to their limits. He had managed not to purge himself of
some seemingly unscientific ideas from eighteenth-century Ro-

manticism. He wanted to do science that would be new. He began by putting aside any thought of understanding real complexity and instead turned to the simplest nonlinear equations he could find.

THE MYSTERY OF THE UNIVERSE first announced itself to the four-year-old Mitchell Feigenbaum through a Silvertone radio sitting in his parents' living room in the Flatbush section of Brooklyn soon after the war. He was dizzy with the thought of music arriving from no tangible cause. The phonograph, on the other hand, he felt he understood. His grandmother had given him a special dispensation to put on the 78s.

His father was a chemist who worked for the Port of New York Authority and later for Clairol. His mother taught in the city's public schools. Mitchell first decided to become an electrical engineer, a sort of professional known in Brooklyn to make a good living. Later he realized that what he wanted to know about a radio was more likely to be found in physics. He was one of a generation of scientists raised in the outer boroughs of New York who made their way to brilliant careers via the great public high schools—in his case, Samuel J. Tilden—and then City College.

Growing up smart in Brooklyn was in some measure a matter of steering an uneven course between the world of mind and the world of other people. He was immensely gregarious when very young, which he regarded as a key to not being beaten up. But something clicked when he realized he could learn things. He became more and more detached from his friends. Ordinary conversation could not hold his interest. Sometime in his last year of college, it struck him that he had missed his adolescence, and he made a deliberate project out of regaining touch with humanity. He would sit silently in the cafeteria, listening to students chatting about shaving or food, and gradually he relearned much of the science of talking to people.

He graduated in 1964 and went on to the Massachusetts Institute of Technology, where he got his doctorate in elementary particle physics in 1970. Then he spent a fruitless four years at Cornell and at the Virginia Polytechnic Institute—fruitless, that is, in terms of the steady publication of work on manageable problems that is essential for a young university scientist. Postdocs were supposed to produce papers. Occasionally an advisor would ask Feigenbaum what had happened to some problem, and he would say, "Oh, I understood it."

Newly installed at Los Alamos, Carruthers, a formidable scientist in his own right, prided himself on his ability to spot talent. He looked not for intelligence but for a sort of creativity that seemed to flow from some magic gland. He always remembered the case of Kenneth Wilson, another soft-spoken Cornell physicist who seemed to be producing absolutely nothing. Anyone who talked to Wilson for long realized that he had a deep capacity for seeing into physics. So the question of Wilson's tenure became a subject of serious debate. The physicists willing to gamble on his unproven potential prevailed—and it was as if a dam burst. Not one but a flood of papers came forth from Wilson's desk drawers, including work that won him the Nobel Prize in 1982.

Wilson's great contribution to physics, along with work by two other physicists, Leo Kadanoff and Michael Fisher, was an important ancestor of chaos theory. These men, working independently, were all thinking in different ways about what happened in phase transitions. They were studying the behavior of matter near the point where it changes from one state to another— from liquid to gas, or from unmagnetized to magnetized. As singular boundaries between two realms of existence, phase transitions tend to be highly nonlinear in their mathematics. The smooth and predictable behavior of matter in any one phase tends to be little help in understanding the transitions. A pot of water on the stove heats up in a regular way until it reaches the boiling point. But then the change in temperature pauses while something quite interesting happens at the molecular interface between liquid and gas.

As Kadanoff viewed the problem in the 1960s, phase transitions pose an intellectual puzzle. Think of a block of metal being magnetized. As it goes into an ordered state, it must make a decision. The magnet can be oriented one way or the other. It is free to choose. But each tiny piece of the metal must make the same choice. How?

Somehow, in the process of choosing, the atoms of the metal

must communicate information to one another. Kadanoff's insight was that the communication can be most simply described in terms of scaling. In effect, he imagined dividing the metal into boxes. Each box communicates with its immediate neighbors. The way to describe that communication is the same as the way to describe the communication of any atom with *its* neighbors. Hence the usefulness of scaling: the best way to think of the metal is in terms of a fractal-like model, with boxes of all different sizes.

Much mathematical analysis, and much experience with real systems, was needed to establish the power of the scaling idea. Kadanoff felt that he had taken an unwieldy business and created a world of extreme beauty and self-containedness. Part of the beauty lay in its universality. Kadanoff's idea gave a backbone to the most striking fact about critical phenomena, namely that these seemingly unrelated transitions—the boiling of liquids, the magnetizing of metals—all follow the same rules.

Then Wilson did the work that brought the whole theory together under the rubric of renormalization group theory, providing a powerful way of carrying out real calculations about real systems. Renormalization had entered physics in the 1940s as a part of quantum theory that made it possible to calculate interactions of electrons and photons. A problem with such calculations, as with the calculations Kadanoff and Wilson worried about, was that some items seemed to require treatment as infinite quantities, a messy and unpleasant business. Renormalizing the system, in ways devised by Richard Feynman, Julian Schwinger, Freeman Dyson, and other physicists, got rid of the infinities.

Only much later, in the 1960s, did Wilson dig down to the underlying basis for renormalization's success. Like Kadanoff, he thought about scaling principles. Certain quantities, such as the mass of a particle, had always been considered fixed—as the mass of any object in everyday experience is fixed. The renormalization shortcut succeeded by acting as though a quantity like mass were not fixed at all. Such quantities seemed to float up or down depending on the scale from which they were viewed. It seemed absurd. Yet it was an exact analogue of what Benoit Mandelbrot was realizing about geometrical shapes and the coastline of England. Their length could not be measured independent of scale. There was a kind of relativity in which the position of the observer, near or far, on the beach or in a satellite, affected the measurement. As Mandelbrot, too, had seen, the variation across scales was not arbitrary; it followed rules. Variability in the standard measures of mass or length meant that a different sort of quantity was remaining fixed. In the case of fractals, it was the fractional dimension—a constant that could be calculated and used as a tool for further calculations. Allowing mass to vary depending on scale meant that mathematicians could recognize similarity across scales.

So for the hard work of calculation, Wilson's renormalization group theory provided a different route into infinitely dense problems. Until then the only way to approach highly nonlinear problems was with a device called perturbation theory. For purposes of calculation, you assume that the nonlinear problem is reasonably close to some solvable, linear problem—just a small perturbation away. You solve the linear problem and perform a complicated bit of trickery with the leftover part, expanding it into what are called Feynman diagrams. The more accuracy you need, the more of these agonizing diagrams you must produce. With luck, your calculations converge toward a solution. Luck has a way of vanishing, however, whenever a problem is especially interesting. Feigenbaum, like every other young particle physicist in the 1960s, found himself doing endless Feynman diagrams. He was left with the conviction that perturbation theory was tedious, nonilluminating, and stupid. So he loved Wilson's new renormalization group theory. By acknowledging self-similarity, it gave a way of collapsing the complexity, one layer at a time.

In practice the renormalization group was far from foolproof. It required a good deal of ingenuity to choose just the right calculations to capture the self-similarity. However, it worked well enough and often enough to inspire some physicists, Feigenbaum included, to try it on the problem of turbulence. After all, self-similarity seemed to be the signature of turbulence, fluctuations upon fluctuations, whorls upon whorls. But what about the onset of turbulence—the mysterious moment when an orderly system turned chaotic. There was no evidence that the renormalization group had anything to say about this transition. There was no evidence, for example, that the transition obeyed laws of scaling.

As a graduate student at M.I.T., Feigenbaum had an experience that stayed with him for many years. He was walking with friends around the Lincoln Reservoir in Boston. He was developing a habit of taking four- and five-hour walks, attuning himself to the panoply of impressions and ideas that would flow through his mind. On this day he became detached from the group and walked alone. He passed some picnickers and, as he moved away, he glanced back every so often, hearing the sounds of their voices, watching the motions of hands gesticulating or reaching for food. Suddenly he felt that the tableau had crossed some threshold into incomprehensibility. The figures were too small to be made out. The actions seemed disconnected, arbitrary, random. What faint sounds reached him had lost meaning.

The ceaseless motion and incomprehensible bustle of life. Feigenbaum recalled the words of Gustav Mahler, describing a sensation that he tried to capture in the third movement of his Second Symphony. *Like the motions of dancing figures in a brilliantly lit ballroom into which you look from the dark night outside and from such a distance that the music is inaudible. . . . Life may appear senseless to you.* Feigenbaum was listening to Mahler and reading Goethe, immersing himself in their high Romantic attitudes. Inevitably it was Goethe's *Faust* he most reveled in, soaking up its combination of the most passionate ideas about the world with the most intellectual. Without some Romantic inclinations, he surely would have dismissed a sensation like his confusion at the reservoir. After all, why shouldn't phenomena lose meaning as they are seen from greater distances? Physical laws provided a trivial explanation for their shrinking. On second thought the connection between shrinking and loss of meaning was not so obvious. Why should it be that as things become small they also become incomprehensible?

He tried quite seriously to analyze this experience in terms of the tools of theoretical physics, wondering what he could say about the brain's machinery of perception. You see some human transactions and you make deductions about them. Given the vast amount of information available to your senses, how does your decoding apparatus sort it out? Clearly—or almost clearly—the brain does not own any direct copies of stuff in the world. There is no library of forms and ideas against which to compare the images of perception. Information is stored in a plastic way, allowing fantastic juxtapositions and leaps of imagination. Some chaos exists out there, and the brain seems to have more flexibility than classical physics in finding the order in it.

At the same time, Feigenbaum was thinking about color. One of the minor skirmishes of science in the first years of the nineteenth century was a difference of opinion between Newton's followers in England and Goethe in Germany over the nature of color. To Newtonian physics, Goethe's ideas were just so much pseudoscientific meandering. Goethe refused to view color as a static quantity, to be measured in a spectrometer and pinned down like a butterfly to cardboard. He argued that color is a matter of perception. "With light poise and counterpoise, Nature oscillates within her prescribed limits," he wrote, "yet thus arise all the varieties and conditions of the phenomena which are presented to us in space and time."

The touchstone of Newton's theory was his famous experiment with a prism. A prism breaks a beam of white light into a rainbow of colors, spread across the whole visible spectrum, and Newton realized that those pure colors must be the elementary components that add to produce white. Further, with a leap of insight, he proposed that the colors corresponded to frequencies. He imagined that some vibrating bodies—corpuscles was the antique word—must be producing colors in proportion to the speed of the vibrations. Considering how little evidence supported this notion, it was as unjustifiable as it was brilliant. What is red? To a physicist, it is light radiating in waves between 620 to 800 billionths of a meter long. Newton's optics proved themselves a thousand times over, while Goethe's treatise on color faded into merciful obscurity. When Feigenbaum went looking for it, he discovered that the one copy in Harvard's libraries had been removed.

He finally did track down a copy, and he found that Goethe had actually performed an extraordinary set of experiments in his investigation of colors. Goethe began as Newton had, with a prism. Newton had held a prism before a light, casting the divided beam onto a white surface. Goethe held the prism to his eye and looked through it. He perceived no color at all, neither a rainbow nor individual hues. Looking at a clear white surface or a clear blue sky through the prism produced the same effect: uniformity.

But if a slight spot interrupted the white surface or a cloud appeared in the sky, then he would see a burst of color. It is "the interchange of light and shadow," Goethe concluded, that causes color. He went on to explore the way people perceive shadows cast by different sources of colored light. He used candles and pencils, mirrors and colored glass, moonlight and sunlight, crystals, liquids, and color wheels in a thorough range of experiments. For example, he lit a candle before a piece of white paper at twilight and held up a pencil. The shadow in the candlelight was a brilliant blue. Why? The white paper alone is perceived as white, either in the declining daylight or in the added light of the warmer candle. How does a shadow divide the white into a region of blue and a region of reddish-yellow? Color is "a degree of darkness," Goethe argued, "allied to shadow." Above all, in a more modern language, color comes from boundary conditions and singularities.

Where Newton was reductionist, Goethe was holistic. Newton broke light apart and found the most basic physical explanation for color. Goethe walked through flower gardens and studied paintings, looking for a grand, all-encompassing explanation. Newton made his theory of color fit a mathematical scheme for all of physics. Goethe, fortunately or unfortunately, abhorred mathematics.

Feigenbaum persuaded himself that Goethe had been right about color. Goethe's ideas resemble a facile notion, popular among psychologists, that makes a distinction between hard physical reality and the variable subjective perception of it. The colors we perceive vary from time to time and from person to person— that much is easy to say. But as Feigenbaum understood them, Goethe's ideas had more true science in them. They were hard and empirical. Over and over again, Goethe emphasized the repeatability of his experiments. It was the perception of color, to Goethe, that was universal and objective. What scientific evidence was there for a definable real-world quality of redness independent of our perception?

Feigenbaum found himself asking what sort of mathematical formalisms might correspond to human perception, particularly a perception that sifted the messy multiplicity of experience and found universal qualities. Redness is not necessarily a particular bandwidth of light, as the Newtonians would have it. It is a territory of a chaotic universe, and the boundaries of that territory are not so easy to describe—yet our minds find redness with regular and verifiable consistency. These were the thoughts of a young physicist, far removed, it seemed, from such problems as fluid turbulence. Still, to understand how the human mind sorts through the chaos of perception, surely one would need to understand how disorder can produce universality.

WHEN FEIGENBAUM BEGAN to think about nonlinearity at Los Alamos, he realized that his education had taught him nothing useful. To solve a system of nonlinear differential equations was impossible, notwithstanding the special examples constructed in textbooks. Perturbative technique, making successive corrections to a solvable problem that one hoped would lie somewhere nearby the real one, seemed foolish. He read through texts on nonlinear flows and oscillations and decided that little existed to help a reasonable physicist. His computational equipment consisting solely of pencil and paper, Feigenbaum decided to start with an analogue of the simple equation that Robert May studied in the context of population biology.

It happened to be the equation high school students use in geometry to graph a parabola. It can be written as $y = r(x - x^2)$. Every value of x produces a value of y, and the resulting curve expresses the relation of the two numbers for the range of values. If x (this year's population) is small, then y (next year's) is small, but larger than x; the curve is rising steeply. If x is in the middle of the range, then y is large. But the parabola levels off and falls, so that if x is large, then y will be small again. That is what produces the equivalent of population crashes in ecological modeling, preventing unrealistic unrestrained growth.

For May and then Feigenbaum, the point was to use this simple calculation not once, but repeated endlessly as a feedback loop. The output of one calculation was fed back in as input for the next. To see what happened graphically, the parabola helped enormously. Pick a starting value along the x axis. Draw a line up to where it meets the parabola. Read the resulting value off the y axis. And start all over with the new value. The sequence bounces from place to place on the parabola at first, and then, perhaps, homes in on a stable equilibrium, where x and y are equal and the value thus does not change.

In spirit, nothing could have been further removed from the complex calculations of standard physics. Instead of a labyrinthine scheme to be solved one time, this was a simple calculation performed over and over again. The numerical experimenter would *watch*, like a chemist peering at a reaction bubbling away inside a beaker. Here the output was just a string of numbers, and it did not always converge to a steady final state. It could end up oscillating back and forth between two values. Or as May had explained to population biologists, it could keep on changing chaotically as long as anyone cared to watch. The choice among these different possible behaviors depended on the value of the tuning parameter.

Feigenbaum carried out numerical work of this faintly experimental sort and, at the same time, tried more traditional theoretical ways of analyzing nonlinear functions. Even so, he could not see the whole picture of what this equation could do. But he could see that the possibilities were already so complicated that they would be viciously hard to analyze. He also knew that three Los Alamos mathematicians—Nicholas Metropolis, Paul Stein, and Myron Stein—had studied such "maps" in 1971, and now Paul Stein warned him that the complexity was frightening indeed. If this simplest of equations already proved intractable, what about the far more complicated equations that a scientist would write down for *real* systems? Feigenbaum put the whole problem on the shelf.

In the brief history of chaos, this one innocent-looking equation provides the most succinct example of how different sorts of scientists looked at one problem in many different ways. To the biologists, it was an equation with a message: Simple systems can do complicated things. To Metropolis, Stein, and Stein, the problem was to catalogue a collection of topological patterns without reference to any numerical values. They would begin the feedback process at a particular point and watch the succeeding values bounce from place to place on the parabola. As the values moved to the right or the left, they wrote down sequences of R's and L's. Pattern number one: R. Pattern number two: RLR. Pattern number 193: RLLLLLRRLL. These sequences had some interesting features to a mathematician—they always seemed to repeat in the same special order. But to a physicist they looked obscure and tedious.

No one realized it then, but Lorenz had looked at the same equation in 1964, as a metaphor for a deep question about climate. The question was so deep that almost no one had thought to ask it before: *Does a climate exist?* That is, does the earth's weather have a long-term average? Most meteorologists, then as now, took the answer for granted. Surely any measurable behavior, no matter how it fluctuates, must have an average. Yet on reflection, it is far from obvious. As Lorenz pointed out, the average weather for the last 12,000 years has been notably different than the average for the previous 12,000, when most of North America was covered by ice. Was there one climate that changed to another for some physical reason? Or is there an even longer-term climate within which those periods were just fluctuations? Or is it possible that a system like the weather may *never* converge to an average?

Lorenz asked a second question. Suppose you could actually write down the complete set of equations that govern the weather. In other words, suppose you had God's own code. Could you then use the equations to calculate average statistics for temperature or rainfall? If the equations were linear, the answer would be an easy yes. But they are nonlinear. Since God has not made the actual equations available, Lorenz instead examined the quadratic difference equation.

Like May, Lorenz first examined what happened as the equation was iterated, given some parameter. With low parameters he saw the equation reaching a stable fixed point. There, certainly, the system produced a "climate" in the most trivial sense possible—the "weather" never changed. With higher parameters he saw the possibility of oscillation between two points, and there, too, the system converged to a simple average. But beyond a certain point, Lorenz saw that chaos ensues. Since he was thinking about climate, he asked not only whether continual feedback would produce periodic behavior, but also what the average output would be. And he recognized that the answer was that the average, too, fluctuated unstably. When the parameter value was changed ever so slightly, the average might change dramatically. By analogy, the earth's climate might never settle reliably into an equilibrium with average long-term behavior.

As a mathematics paper, Lorenz's climate work would have been a failure—he proved nothing in the axiomatic sense. As a physics paper, too, it was seriously flawed, because he could not justify using such a simple equation to draw conclusions about the earth's climate. Lorenz knew what he was saying, though. "The writer feels that this resemblance is no mere accident, but that the difference equation captures much of the mathematics, even if not the physics, of the transitions from one regime of flow to another, and, indeed, of the whole phenomenon of instability." Even twenty years later, no one could understand what intuition justified such a bold claim, published in *Tellus*, a Swedish meteorology journal. ("*Tellus!* Nobody reads *Tellus*," a physicist exclaimed bitterly.) Lorenz was coming to understand ever more deeply the peculiar possibilities of chaotic systems—more deeply than he could express in the language of meteorology.

As he continued to explore the changing masks of dynamical systems, Lorenz realized that systems slightly more complicated than the quadratic map could produce other kinds of unexpected patterns. Hiding within a particular system could be more than one stable solution. An observer might see one kind of behavior over a very long time, yet a completely different kind of behavior could be just as natural for the system. Such a system is called intransitive. It can stay in one equilibrium or the other, but not both. Only a kick from outside can force it to change states. In a trivial way, a standard pendulum clock is an intransitive system. A steady flow of energy comes in from a wind-up spring or a battery through an escapement mechanism. A steady flow of energy is drained out by friction. The obvious equilibrium state is a regular swinging motion. If a passerby bumps the clock, the pendulum might speed up or slow down from the momentary jolt but will quickly return to its equilibrium. But the clock has a second equilibrium as well—a second valid solution to its equations of motion—and that is the state in which the pendulum is hanging straight down and not moving. A less trivial intransitive system—perhaps with several distinct regions of utterly different behavior—could be climate itself.

Climatologists who use global computer models to simulate the long-term behavior of the earth's atmosphere and oceans have known for several years that their models allow at least one dramatically different equilibrium. During the entire geological past, this alternative climate has never existed, but it could be an equally valid solution to the system of equations governing the earth. It is what some climatologists call the White Earth climate: an earth whose continents are covered by snow and whose oceans are covered by ice. A glaciated earth would reflect seventy percent of the incoming solar radiation and so would stay extremely cold. The lowest layer of the atmosphere, the troposphere, would be much thinner. The storms that would blow across the frozen surface would be much smaller than the storms we know. In general, the climate would be less hospitable to life as we know it. Computer models have such a strong tendency to fall into the White Earth equilibrium that climatologists find themselves wondering why it has never come about. It may simply be a matter of chance.

To push the earth's climate into the glaciated state would require a huge kick from some external source. But Lorenz described yet another plausible kind of behavior called "almost-intransitivity." An almost-intransitive system displays one sort of average behavior for a very long time, fluctuating within certain bounds. Then, for no reason whatsoever, it shifts into a different sort of behavior, still fluctuating but producing a different average. The people who design computer models are aware of Lorenz's discovery, but they try at all costs to avoid almost-intransitivity. It is too unpredictable. Their natural bias is to make models with a strong tendency to return to the equilibrium we measure every day on the real planet. Then, to explain large changes in climate, they look for external causes—changes in the earth's orbit around the sun, for example. Yet it takes no great imagination for a climatologist to see that almost-intransitivity might well explain why the earth's climate has drifted in and out of long Ice Ages at mysterious, irregular intervals. If so, no physical cause need be found for the timing. The Ice Ages may simply be a byproduct of chaos.

LIKE A GUN COLLECTOR wistfully recalling the Colt .45 in the era of automatic weaponry, the modern scientist nurses a certain nostalgia for the HP-65 hand-held calculator. In the few years of its supremacy, this machine changed many scientists' working habits forever. For Feigenbaum, it was the bridge between pencil-and-paper and a style of working with computers that had not yet been conceived.

He knew nothing of Lorenz, but in the summer of 1975, at a gathering in Aspen, Colorado, he heard Steve Smale talk about some of the mathematical qualities of the same quadratic difference equation. Smale seemed to think that there were some interesting open questions about the exact point at which the mapping changes from periodic to chaotic. As always, Smale had a sharp instinct for questions worth exploring. Feigenbaum decided to look into it once more. With his calculator he began to use a combination of analytic algebra and numerical exploration to piece together an understanding of the quadratic map, concentrating on the boundary region between order and chaos.

Metaphorically—but *only* metaphorically—he knew that this region was like the mysterious boundary between smooth flow and turbulence in a fluid. It was the region that Robert May had called to the attention of population biologists who had previously failed to notice the possibility of any but orderly cycles in changing animal populations. En route to chaos in this region was a cascade of period-doublings, the splitting of two-cycles into four-cycles, four-cycles into eight-cycles, and so on. These splittings made a a fascinating pattern. They were the points at which a slight change in *fecundity*, for example, might lead a population of gypsy moths to change from a four-year cycle to an eight-year cycle. Feigenbaum decided to begin by calculating the exact parameter values that produced the splittings.

In the end, it was the slowness of the calculator that led him to a discovery that August. It took ages—minutes, in fact—to calculate the exact parameter value of each period-doubling. The higher up the chain he went, the longer it took. With a fast computer, and with a printout, Feigenbaum might have observed no pattern. But he had to write the numbers down by hand, and then he had to think about them while he was waiting, and then, to save time, he had to guess where the next answer would be.

Yet all in an instant he saw that he did not have to guess. There was an unexpected regularity hidden in this system: the numbers were converging geometrically, the way a line of identical telephone poles converges toward the horizon in a perspective drawing. If you know how big to make any two telephone poles, you know all the rest; the ratio of the second to the first will also be the ratio of the third to the second, and so on. The period-doublings were not just coming faster and faster, but they were coming faster and faster at a constant rate.

Why should this be so? Ordinarily, the presence of geometric convergence suggests that something, somewhere, is repeating itself on different scales. But if there was a scaling pattern inside this equation, no one had ever seen it. Feigenbaum calculated the ratio of convergence to the finest precision possible on his machine—three decimal places—and came up with a number, 4.669. Did this particular ratio mean anything? Feigenbaum did what anyone would do who cared about numbers. He spent the rest of the day trying to fit the number to all the standard constants—π, e, and so forth. It was a variant of none.

Oddly, Robert May realized later that he, too, had seen this geometric convergence. But he forgot it as quickly as he noted it. From May's perspective in ecology, it was a numerical peculiarity and nothing more. In the real-world systems he was considering, systems of animal populations or even economic models, the inevitable noise would overwhelm any detail that precise. The very messiness that had led him so far stopped him at the crucial point. May was excited by the gross behavior of the equation. He never imagined that the numerical details would prove important.

Feigenbaum knew what he had, because geometric convergence meant that something in this equation was *scaling,* and he knew that scaling was important. All of renormalization theory depended on it. In an apparently unruly system, scaling meant that some quality was being preserved while everything else changed. Some regularity lay beneath the turbulent surface of the equation. But where? It was hard to see what to do next.

Summer turns rapidly to autumn in the rarefied Los Alamos air, and October had nearly ended when Feigenbaum was struck by an odd thought. He knew that Metropolis, Stein, and Stein had looked at other equations as well and had found that certain patterns carried over from one sort of function to another. The same combinations of R's and L's appeared, and they appeared in the same order. One function had involved the sine of a number, a twist that made Feigenbaum's carefully worked-out approach to the parabola equation irrelevant. He would have to start over. So he took his HP-65 again and began to compute the period-doublings for $x_{t+1} = r \sin \pi x_t$. Calculating a trigonometric function made the process that much slower, and Feigenbaum wondered whether, as with the simpler version of the equation, he would be able to use a shortcut. Sure enough, scanning the numbers, he realized that they were again converging geometrically. It was simply a matter of calculating the convergence rate for this new equation. Again, his precision was limited, but he got a result to three decimal places: 4.669.

It was the same number. Incredibly, this trigonometric function was not just displaying a consistent, geometric regularity. It was displaying a regularity that was numerically *identical* to that of a much simpler function. No mathematical or physical theory existed to explain why two equations so different in form and meaning should lead to the same result.

Feigenbaum called Paul Stein. Stein was not prepared to believe the coincidence on such scanty evidence. The precision was low, after all. Nevertheless, Feigenbaum also called his parents in New Jersey to tell them he had stumbled across something profound. He told his mother it was going to make him famous. Then he started trying other functions, anything he could think of that went through a sequence of bifurcations on the way to disorder. Every one produced the same number.

Feigenbaum had played with numbers all his life. When he was a teen-ager he knew how to calculate logarithms and sines that most people would look up in tables. But he had never learned to use any computer bigger than his hand calculator—and in this he was typical of physicists and mathematicians, who tended to disdain the mechanistic thinking that computer work implied. Now, though, it was time. He asked a colleague to teach him Fortran, and, by the end of the day, for a variety of functions, he had calculated his constant to five decimal places, 4.66920. That night he read about double precision in the manual, and the next day he got as far as 4.6692016090—enough precision to convince Stein. Feigenbaum wasn't quite sure he had convinced himself, though. He had set out to look for regularity—that was what understanding mathematics meant—but he had also set out *knowing* that particular kinds of equations, just like particular physical systems, behave in special, characteristic ways. These equations were simple, after all. Feigenbaum understood the quadratic equation, he understood the sine equation—the mathematics was trivial. Yet something in the heart of these very different equations, repeating over and over again, created a single number. He had stumbled upon something: perhaps just a curiosity; perhaps a new law of nature.

Imagine that a prehistoric zoologist decides that some things are heavier than other things—they have some abstract quality he calls *weight*—and he wants to investigate this idea scientifically. He has never actually measured weight, but he thinks he has some understanding of the idea. He looks at big snakes and little snakes, big bears and little bears, and he guesses that the weight of these animals might have some relationship to their size. He builds a scale and starts weighing snakes. To his astonishment, every snake weighs the same. To his consternation, every bear weighs the same, too. And to his further amazement, bears weigh the same as snakes. They all weigh 4.6692016090. Clearly *weight* is not what he supposed. The whole concept requires rethinking.

Rolling streams, swinging pendulums, electronic oscillators—many physical systems went through a transition on the way to chaos, and those transitions had remained too complicated for analysis. These were all systems whose mechanics seemed perfectly well understood. Physicists knew all the right equations; yet moving from the equations to an understanding of global, long-term behavior seemed impossible. Unfortunately, equations for fluids, even pendulums, were far more challenging than the simple one-dimensional logistic map. But Feigenbaum's discovery implied that those equations were beside the point. They were irrelevant. When order emerged, it suddenly seemed to have forgotten what the original equation was. Quadratic or trigonometric, the result was the same. "The whole tradition of physics is that you isolate the mechanisms and then all the rest flows," he said. "That's completely falling apart. Here you know the right equations but they're just not helpful. You add up all the microscopic pieces and you find that you cannot extend them to the long term. They're not what's important in the problem. It completely changes what it means to *know* something."

Although the connection between numerics and physics was faint, Feigenbaum had found evidence that he needed to work out a new way of calculating complex nonlinear problems. So far, all available techniques had depended on the details of the functions. If the function was a sine function, Feigenbaum's carefully worked-out calculations were sine calculations. His discovery of universality meant that all those techniques would have to be thrown out. The regularity had nothing to do with sines. It had nothing to do with parabolas. It had nothing to do with any particular function. But why? It was frustrating. Nature had pulled back a curtain for an instant and offered a glimpse of unexpected order. What else was behind that curtain?

WHEN INSPIRATION CAME, it was in the form of a picture, a mental image of two small wavy forms and one big one. That was all—a bright, sharp image etched in his mind, no more, perhaps, than the visible top of a vast iceberg of mental processing that had taken place below the waterline of consciousness. It had to do with scaling, and it gave Feigenbaum the path he needed.

He was studying attractors. The steady equilibrium reached by his mappings is a fixed point that attracts all others—no matter what the starting "population," it will bounce steadily in toward the attractor. Then, with the first period-doubling, the attractor splits in two, like a dividing cell. At first, these two points are practically together; then, as the parameter rises, they float apart. Then another period-doubling: each point of the attractor divides again, at the same moment. Feigenbaum's number let him predict *when* the period-doublings would occur. Now he discovered that he could also predict the precise values of each point on this ever-more-complicated attractor—two points, four points, eight points . . . He could predict the actual populations reached in the year-to-year oscillations. There was yet another geometric convergence. These numbers, too, obeyed a law of scaling.

Feigenbaum was exploring a forgotten middle ground be-

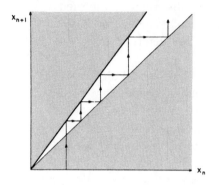

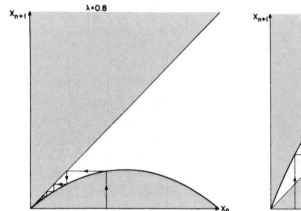

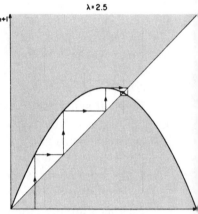

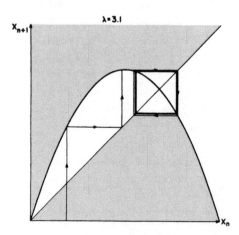

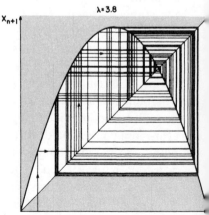

H. Bruce Stewart, J. M. Thompson / Nancy Sterngold

ZEROING IN ON CHAOS. A simple equation, repeated many times over: Mitchell Feigenbaum focused on straightforward functions, taking one number as input and producing another as output. For animal populations, a function might express the relationship between this year's population and next year's.

One way to visualize such functions is to make a graph, plotting input on the horizontal axis and output on the vertical axis. For each possible input, x, there is just one output, y, and these form a shape represented by the heavy line.

Then, to represent the long-term behavior of the system, Feigenbaum drew a trajectory that started with some arbitrary x. Because each y was then fed back into the same function as new input, he could use a sort of schematic shortcut: The trajectory would bounce off the 45-degree line, the line where x equals y.

For an ecologist, the most obvious sort of function for population growth is linear—the Malthusian scenario of steady, limitless growth by a fixed percentage each year (left). More realistic functions formed an arch, sending the population back downward when it became too high. Illustrated is the "logistic map," a perfect parabola, defined by the function $y = rx(1 - x)$, where the value of r, from 0 to 4, determines the parabola's steepness. But Feigenbaum discovered that it did not matter precisely what sort of arch he used; the details of the equation were beside the point. What mattered was that the function should have a "hump."

The behavior depended sensitively, though, on the steepness—the degree of nonlinearity, or what Robert May called "boom-and-bustiness." Too shallow a function would produce extinction: Any starting population would lead eventually to zero. Increasing the steepness produced the steady equilibrium that a traditional ecologist would expect; that point, drawing in all trajectories, was a one-dimensional "attractor."

Beyond a certain point, a bifurcation produced an oscillating population with period two. Then more period-doublings would occur, and finally (bottom right) the trajectory would refuse to settle down at all.

Such images were a starting point for Feigenbaum when he tried to construct a theory. He began thinking in terms of recursion: functions of functions, and functions of functions of functions, and so on; maps with two humps, and then four. . . .

tween mathematics and physics. His work was hard to classify. It was not mathematics; he was not proving anything. He was studying numbers, yes, but numbers are to a mathematician what bags of coins are to an investment banker: nominally the stuff of his profession, but actually too gritty and particular to waste time on. Ideas are the real currency of mathematicians. Feigenbaum was carrying out a program in physics, and, strange as it seemed, it was almost a kind of experimental physics.

Numbers and functions were his object of study, instead of mesons and quarks. They had trajectories and orbits. He needed to inquire into their behavior. He needed—in a phrase that later became a cliché of the new science—to *create intuition*. His accelerator and his cloud chamber were the computer. Along with his theory, he was building a methodology. Ordinarily a computer user would construct a problem, feed it in, and wait for the machine to calculate its solution—one problem, one solution. Feigenbaum and the chaos researchers who followed needed more. They needed to do what Lorenz had done, to create miniature universes and observe their evolution. Then they could change this feature or that and observe the changed paths that would result. They were armed with the new conviction, after all, that tiny changes in certain features could lead to remarkable changes in overall behavior.

Feigenbaum quickly discovered how ill-suited the computer facilities of Los Alamos were for the style of computing he wanted to develop. Despite enormous resources, far greater than at most universities, Los Alamos had few terminals capable of displaying graphs and pictures, and those few were in the Weapons Division. Feigenbaum wanted to take numbers and plot them as points on a map. He had to resort to the most primitive method conceivable: long rolls of printout paper with lines made by printing rows of spaces followed by an asterisk or a plus sign. The official policy at Los Alamos held that one big computer was worth far more than many little computers—a policy that went with the *one problem, one solution* tradition. Little computers were discouraged. Furthermore, any division's purchase of a computer would have to meet stringent government guidelines and a formal review. Only later, with the budgetary complicity of the Theoretical Division, did Feigenbaum become the recipient of a $20,000 "desktop calculator." Then he could change his equations and pictures on the run, tweaking them and tuning them, playing the computer like a musical instrument. For now, the only terminals capable of serious graphics were in high-security areas—behind the fence, in local parlance. Feigenbaum had to use a terminal hooked up by telephone lines to a central computer. The reality of working in such an arrangement made it hard to appreciate the raw power of the computer at the other end of the line. Even the simplest tasks took minutes. To edit a line of a program meant pressing *Return* and waiting while the terminal hummed incessantly and the central computer played its electronic round robin with other users across the laboratory.

While he was computing, he was thinking. What new mathematics could produce the multiple scaling patterns he was observing? Something about these functions must be *recursive*, he realized, *self-referential*, the behavior of one guided by the behavior of another hidden inside it. The wavy image that had come to him in a moment of inspiration expressed something about the way one function could be scaled to match another. He applied the mathematics of renormalization group theory, with its use of scaling to collapse infinities into manageable quantities. In the spring of 1976 he entered a mode of existence more intense than any he had lived through. He would concentrate as if in a trance, programming furiously, scribbling with his pencil, programming again. He could not call C division for help, because that would mean signing off the computer to use the telephone, and reconnection was chancy. He could not stop for more than five minutes' thought, because the computer would automatically disconnect his line. Every so often the computer would go down anyway, leaving him shaking with adrenalin. He worked for two months without pause. His functional day was twenty-two hours. He would try to go to sleep in a kind of buzz, and awaken two hours later with his thoughts exactly where he had left them. His diet was strictly coffee. (Even when healthy and at peace, Feigenbaum subsisted exclusively on the reddest possible meat, coffee, and red wine. His friends speculated that he must be getting his vitamins from cigarettes.)

In the end, a doctor called it off. He prescribed a modest regimen of Valium and an enforced vacation. But by then Feigenbaum had created a universal theory.

UNIVERSALITY MADE THE DIFFERENCE between beautiful and useful. Mathematicians, beyond a certain point, care little whether they are providing a technique for calculation. Physicists, beyond a certain point, need numbers. Universality offered the hope that by solving an easy problem physicists could solve much harder problems. The answers would be the same. Further, by placing his theory in the framework of the renormalization group, Feigenbaum gave it a clothing that physicists would recognize as a tool for calculating, almost something standard.

But what made universality useful also made it hard for physicists to believe. Universality meant that different systems would behave identically. Of course, Feigenbaum was only studying simple numerical functions. But he believed that his theory expressed a natural law about systems at the point of transition between orderly and turbulent. Everyone knew that turbulence meant a continuous spectrum of different frequencies, and everyone had wondered where the different frequencies came from. Suddenly you could see the frequencies coming in sequentially. The physical implication was that real-world systems would behave in the same, recognizable way, and that furthermore it would be *measurably* the same. Feigenbaum's universality was not just qualitative, it was quantitative; not just structural, but metrical. It extended not just to patterns, but to precise numbers. To a physicist, that strained credulity.

Years later Feigenbaum still kept in a desk drawer, where he could get at them quickly, his rejection letters. By then he had all the recognition he needed. His Los Alamos work had won him prizes and awards that brought prestige and money. But it still rankled that editors of the top academic journals had deemed his work unfit for publication for two years after he began submitting it. The notion of a scientific breakthrough so original and unexpected that it cannot be published seems a slightly tarnished myth. Modern science, with its vast flow of information and its impartial system of peer review, is not supposed to be a matter of taste. One editor who sent back a Feigenbaum manuscript recognized years later that he had rejected a paper that was a turning point for the field; yet he still argued that the paper had been unsuited to his journal's audience of applied mathematicians. In the meantime, even without publication, Feigenbaum's breakthrough became a superheated piece of news in certain circles of mathematics and physics. The kernel of theory was disseminated the way most science is now disseminated—through lectures and preprints. Feigenbaum described his work at conferences, and requests for photocopies of his papers came in by the score and then by the hundred.

MODERN ECONOMICS RELIES HEAVILY on the efficient market theory. Knowledge is assumed to flow freely from place to place. The people making important decisions are supposed to have access to more or less the same body of information. Of course, pockets of ignorance or inside information remain here and there, but on the whole, once knowledge is public, economists assume that it is known everywhere. Historians of science often take for granted an efficient market theory of their own. When a discovery is made, when an idea is expressed, it is assumed to become the common property of the scientific world. Each discovery and each new insight builds on the last. Science rises like a building, brick by brick. Intellectual chronicles can be, for all practical purposes, linear.

That view of science works best when a well-defined discipline awaits the resolution of a well-defined problem. No one misunderstood the discovery of the molecular structure of DNA, for example. But the history of ideas is not always so neat. As nonlinear science arose in odd corners of different disciplines, the flow of ideas failed to follow the standard logic of historians. The emergence of chaos as an entity unto itself was a story not only of new theories and new discoveries, but also of the belated understanding of old ideas. Many pieces of the puzzle had been seen long before—by Poincaré, by Maxwell, even by Einstein— and then forgotten. Many new pieces were understood at first only by a few insiders. A mathematical discovery was understood by mathematicians, a physics discovery by physicists, a meteorological discovery by no one. The way ideas spread became as important as the way they originated.

Each scientist had a private constellation of intellectual parents. Each had his own picture of the landscape of ideas, and each picture was limited in its own way. Knowledge was imperfect. Scientists were biased by the customs of their disciplines or by the accidental paths of their own educations. The scientific world can be surprisingly finite. No committee of scientists pushed history into a new channel—a handful of individuals did it, with individual perceptions and individual goals.

Afterwards, a consensus began to take shape about which innovations and which contributions had been most influential. But the consensus involved a certain element of revisionism. In the heat of discovery, particularly during the late 1970s, no two physicists, no two mathematicians understood chaos in exactly the same way. A scientist accustomed to classical systems without friction or dissipation would place himself in a lineage descending from Russians like A. N. Kolmogorov and V. I. Arnold. A mathematician accustomed to classical dynamical systems would envision a line from Poincaré to Birkhoff to Levinson to Smale. Later, a mathematician's constellation might center on Smale, Guckenheimer, and Ruelle. Or it might emphasize a computationally inclined set of forebears associated with Los Alamos: Ulam, Metropolis, Stein. A theoretical physicist might think of Ruelle, Lorenz, Rössler, and Yorke. A biologist would think of Smale, Guckenheimer, May, and Yorke. The possible combinations were endless. A scientist working with materials—a geologist or a seismologist—would credit the direct influence of Mandelbrot; a theoretical physicist would barely acknowledge knowing the name.

Feigenbaum's role would become a special source of contention. Much later, when he was riding a crest of semicelebrity, some physicists went out of their way to cite other people who had been working on the same problem at the same time, give or take a few years. Some accused him of focusing too narrowly on a small piece of the broad spectrum of chaotic behavior. "Feigenbaumology" was overrated, a physicist might say—a beautiful piece of work, to be sure, but not as broadly influential as Yorke's work, for example. In 1984, Feigenbaum was invited to address the Nobel Symposium in Sweden, and there the controversy swirled.

Benoit Mandelbrot gave a wickedly pointed talk that listeners later described as his "antifeigenbaum lecture." Somehow Mandelbrot had exhumed a twenty-year-old paper on period-doubling by a Finnish mathematician named Myrberg, and he kept describing the Feigenbaum sequences as "Myrberg sequences." But Feigenbaum had discovered universality and created a theory to explain it. That was the pivot on which the new science swung. Unable to publish such an astonishing and counterintuitive result, he spread the word in a series of lectures at a New Hampshire conference in August 1976, an international mathematics meeting at Los Alamos in September, a set of talks at Brown University in November. The discovery and the theory met surprise, disbelief, and excitement. The more a scientist had thought about nonlinearity, the more he felt the force of Feigenbaum's universality. One put it simply: "It was a very happy and shocking discovery that there were structures in nonlinear systems that are always the same if you looked at them the right way." Some physicists picked up not just the ideas but also the techniques. Playing with these maps—just playing—gave them chills. With their own calculators, they could experience the surprise and satisfaction that had kept Feigenbaum going at Los Alamos. And they refined the theory. Hearing his talk at the Institute for Advanced Study in Princeton, Predrag Cvitanović, a particle physicist, helped Feigenbaum simplify his theory and extend its universality. But all the while, Cvitanović pretended it was just a pastime; he could not bring himself to admit to his colleagues what he was doing.

Among mathematicians, too, a reserved attitude prevailed, largely because Feigenbaum did not provide a rigorous proof. Indeed, not until 1979 did proof come on mathematicians' terms, in work by Oscar E. Lanford III. Feigenbaum often recalled presenting his theory to a distinguished audience at the Los Alamos meeting in September. He had barely begun to describe the work when the eminent mathematician Mark Kac rose to ask: "Sir, do you mean to offer numerics or a proof?"

More than the one and less than the other, Feigenbaum replied.

"Is it what any *reasonable* man would call a proof?"

Feigenbaum said that the listeners would have to judge for themselves. After he was done speaking, he polled Kac, who responded, with a sardonically trilled r: "Yes, that's indeed a reasonable man's proof. The details can be left to the r-r-rigorous mathematicians."

A movement had begun, and the discovery of universality spurred it forward. In the summer of 1977, two physicists, Joseph Ford and Giulio Casati, organized the first conference on a science called chaos. It was held in a gracious villa in Como, Italy, a tiny city at the southern foot of the lake of the same name, a stunningly deep blue catchbasin for the melting snow from the Italian Alps. One hundred people came—mostly physicists, but also curious scientists from other fields. "Mitch had seen universality and found out how it scaled and worked out a way of getting to chaos that was intuitively appealing," Ford said. "It was the first time we had a clear model that everybody could understand.

"And it was one of those things whose time had come. In disciplines from astronomy to zoology, people were doing the same things, publishing in their narrow disciplinary journals, just totally unaware that the other people were around. They thought they were by themselves, and they were regarded as a bit eccentric in their own areas. They had exhausted the simple questions you could ask and begun to worry about phenomena that were a bit more complicated. And these people were just weepingly grateful to find out that everybody else was there, too."

LATER, FEIGENBAUM LIVED in a bare space, a bed in one room, a computer in another, and, in the third, three black electronic towers for playing his solidly Germanic record collection. His one experiment in home furnishing, the purchase of an expensive marble coffee table while he was in Italy, had ended in failure; he received a parcel of marble chips. Piles of papers and books lined the walls. He talked rapidly, his long hair, gray now mixed with brown, sweeping back from his forehead. "Something dramatic happened in the twenties. For no good reason physicists stumbled upon an essentially correct description of the world around them—because the theory of quantum mechanics *is* in some sense essentially correct. It tells you how you can take dirt and make computers from it. It's the way we've learned to manipulate our universe. It's the way chemicals are made and plastics and what not. One knows how to compute with it. It's an extravagantly good theory—except at some level it doesn't make good sense.

"Some part of the imagery is missing. If you ask what the equations really mean and what is the description of the world according to this theory, it's not a description that entails your intuition of the world. You can't think of a particle moving as though it has a trajectory. You're not allowed to visualize it that way. If you start asking more and more subtle questions—what does this theory tell you the world looks like?—in the end it's so far out of your normal way of picturing things that you run into all sorts of conflicts. Now maybe that's the way the world really is. But you don't really know that there isn't another way of assembling all this information that doesn't demand so radical a departure from the way in which you intuit things.

"There's a fundamental presumption in physics that the way you understand the world is that you keep isolating its ingredients until you understand the stuff that you think is truly fundamental. Then you presume that the other things you don't understand are details. The assumption is that there are a small number of principles that you can discern by looking at things in their pure state—this is the true analytic notion—and then somehow you put these together in more complicated ways when you want to solve more dirty problems. If you *can*.

"In the end, to understand you have to change gears. You have to reassemble how you conceive of the important things that are going on. You could have tried to simulate a model fluid system on a computer. It's just beginning to be possible. But it would have been a waste of effort, because what *really* happens has nothing to do with a fluid or a particular equation. It's a general description of what happens in a large variety of systems when things work on themselves again and again. It requires a different way of thinking about the problem.

"When you look at this room—you see junk sitting over there and a person sitting over here and doors over there—you're supposed to take the elementary principles of matter and write down the wave functions to describe them. Well, this is not a feasible thought. Maybe God could do it, but no analytic thought exists for understanding such a problem.

"It's not an academic question any more to ask what's going to happen to a cloud. People very much want to know—and that means there's money available for it. That problem is very much within the realm of physics and it's a problem very much of the same caliber. You're looking at something complicated, and the present way of solving it is to try to look at as many points as you can, enough stuff to say where the cloud is, where the warm air is, what its velocity is, and so forth. Then you stick it into the biggest machine you can afford and you try to get an estimate of what it's going to do next. But this is not very realistic."

He stubbed out one cigarette and lit another. "One has to look for different ways. One has to look for scaling structures—how do big details relate to little details. You look at fluid disturbances, complicated structures in which the complexity has come about by a persistent process. At some level they don't care very much what the size of the process is—it could be the size of a pea or the size of a basketball. The process doesn't care where it is, and moreover it doesn't care how long it's been going. The only things that can ever be universal, in a sense, are scaling things.

"In a way, art is a theory about the way the world looks to human beings. It's abundantly obvious that one doesn't know the world around us in detail. What artists have accomplished is realizing that there's only a small amount of stuff that's important, and then seeing what it was. So they can do some of my research for me. When you look at early stuff of Van Gogh there are zillions of details that are put into it, there's always an immense amount of information in his paintings. It obviously occurred to him, what is the irreducible amount of this stuff that you have to put in. Or you can study the horizons in Dutch ink drawings from around 1600, with tiny trees and cows that look very real. If you look closely, the trees have sort of leafy boundaries, but it doesn't work if that's all it is—there are also, sticking in it, little pieces of twiglike stuff. There's a definite interplay between the softer textures and the things with more definite lines. Somehow the combination gives the correct perception. With Ruysdael and Turner, if you look at the way they construct complicated water, it is clearly done in an iterative way. There's some level of stuff, and then stuff painted on top of that, and then corrections to *that*.

Turbulent fluids for those painters is always something with a scale idea in it.

"I truly do want to know how to describe clouds. But to say there's a piece over here with that much density, and next to it a piece with this much density—to accumulate that much detailed information, I think is wrong. It's certainly not how a human being perceives those things, and it's not how an artist perceives them. Somewhere the business of writing down partial differential equations is not to have done the work on the problem.

"Somehow the wondrous promise of the earth is that there are things beautiful in it, things wondrous and alluring, and by virtue of your trade you want to understand them." He put the cigarette down. Smoke rose from the ashtray, first in a thin column and then (with a nod to universality) in broken tendrils that swirled upward to the ceiling.

The Experimenter

It's an experience like no other experience I can describe, the best thing that can happen to a scientist, realizing that something that's happened in his or her mind exactly corresponds to something that happens in nature. It's startling every time it occurs. One is surprised that a construct of one's own mind can actually be realized in the honest-to-goodness world out there. A great shock, and a great, great joy.

—LEO KADANOFF

"ALBERT IS GETTING MATURE." So they said at École Normale Supérieure, the academy which, with École Polytechnique, sits atop the French educational hierarchy. They wondered whether age was taking its toll on Albert Libchaber, who had made a distinguished name for himself as a low-temperature physicist, studying the quantum behavior of superfluid helium at temperatures a breath away from absolute zero. He had prestige and a secure place on the faculty. And now in 1977 he was wasting his time and the university's resources on an experiment that seemed trivial. Libchaber himself worried that he would be jeopardizing the career of any graduate student he employed on such a project, so he got the assistance of a professional engineer instead.

Five years before the Germans invaded Paris, Libchaber was born there, the son of Polish Jews, the grandson of a rabbi. He survived the war the same way Benoit Mandelbrot did, by hiding in the countryside, separated from his parents because their accents were too dangerous. His parents managed to survive; the rest of the family was lost to the Nazis. In a quirk of political fate, Libchaber's own life was saved by the protection of a local chief of the Pétain secret police, a man whose fervent right-wing beliefs were matched only by his fervent antiracism. After the war, the ten-year-old boy returned the favor. He testified, only half-comprehending, before a war crimes commission, and his testimony saved the man.

Moving through the world of French academic science, Libchaber rose in his profession, his brilliance never questioned. His colleagues did sometimes think he was a little crazy—a Jewish mystic amid the rationalists, a Gaullist where most scientists were Communists. They joked about his Great Man theory of history, his fixation on Goethe, his obsession with old books. He had hundreds of original editions of works by scientists, some dating back to the 1600s. He read them not as historical curiosities but as a source of fresh ideas about the nature of reality, the same reality he was probing with his lasers and his high-technology refrigeration coils. In his engineer, Jean Maurer, he had found a compatible spirit, a Frenchman who worked only when he felt like it. Libchaber thought Maurer would find his new project *amusing*—his understated Gallic euphemism for *intriguing* or *exciting* or *profound*. The two set out in 1977 to build an experiment that would reveal the onset of turbulence.

As an experimenter, Libchaber was known for a nineteenth-century style: clever mind, nimble hands, always preferring ingenuity to brute force. He disliked giant technology and heavy computation. His idea of a good experiment was like a mathematician's idea of a good proof. Elegance counted as much as results. Even so, some colleagues thought he was carrying things too far with his onset-of-turbulence experiment. It was small enough to carry around in a matchbox—and sometimes Libchaber did carry it around, like some piece of conceptual art. He called it "Helium in a Small Box." The heart of the experiment was even smaller, a cell about the size of a lemon seed, carved in stainless steel with the sharpest possible edges and walls. Into the cell was fed liquid helium chilled to about four degrees above absolute zero, warm compared to Libchaber's old superfluid experiments.

The laboratory occupied the second floor of the École physics building in Paris, just a few hundred feet from Louis Pasteur's old laboratory. Like all good general-purpose physics laboratories, Libchaber's existed in a state of constant mess, paint cans and hand tools strewn about on floors and tables, odd-sized pieces of metal and plastic everywhere. Amid the disarray, the apparatus that held Libchaber's minuscule fluid cell was a striking bit of purposefulness. Below the stainless steel cell sat a bottom plate of high-purity copper. Above sat a top plate of sapphire crystal. The materials were chosen according to how they conducted heat.

There were tiny electric heating coils and Teflon gaskets. The liquid helium flowed down from a reservoir, itself just a half-inch cube. The whole system sat inside a container that maintained an extreme vacuum. And that container, in turn, sat in a bath of liquid nitrogen, to help stabilize the temperature.

Vibration always worried Libchaber. Experiments, like real nonlinear systems, existed against a constant background of noise. Noise hampered measurement and corrupted data. In sensitive flows—and Libchaber's would be as sensitive as he could make it—noise might sharply perturb a nonlinear flow, knocking it from one kind of behavior into another. But nonlinearity can stabilize a system as well as destabilize it. Nonlinear feedback regulates motion, making it more robust. In a linear system, a perturbation has a constant effect. In the presence of nonlinearity, a perturbation can feed on itself until it dies away and the system returns

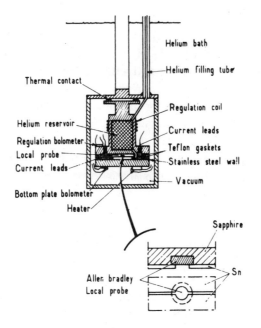

"HELIUM IN A SMALL BOX." Albert Libchaber's delicate experiment: Its heart was a carefully machined rectangular cell containing liquid helium; tiny sapphire "bolometers" measured the fluid's temperature. The tiny cell was embedded in a casing designed to shield it from the noise and vibration and to allow precise control of the heating.

automatically to a stable state. Libchaber believed that biological systems used their nonlinearity as a defense against noise. The transfer of energy by proteins, the wave motion of the heart's electricity, the nervous system—all these kept their versatility in a noisy world. Libchaber hoped that whatever structure underlay fluid flow would prove robust enough for his experiment to detect.

His plan was to create convection in the liquid helium by making the bottom plate warmer than the top plate. It was exactly the convection model described by Edward Lorenz, the classic system known as Rayleigh-Bénard convection. Libchaber was not aware of Lorenz—not yet. Nor had he any idea of Mitchell Feigenbaum's theory. In 1977 Feigenbaum was beginning to travel the scientific lecture circuit, and his discoveries were making their mark where scientists knew how to interpret them. But as far as most physicists could tell, the patterns and regularities of Feigenbaumology bore no obvious connection to real systems. Those patterns came out of a digital calculator. Physical systems were infinitely more complicated. Without more evidence, the most anyone could say was that Feigenbaum had discovered a mathematical analogy that *looked* like the beginning of turbulence.

Libchaber knew that American and French experiments had weakened the Landau idea for the onset of turbulence by showing that turbulence arrived in a sudden transition, instead of a continuous piling-up of different frequencies. Experimenters like Jerry Gollub and Harry Swinney, with their flow in a rotating cylinder, had demonstrated that a new theory was needed, but they had not been able to see the transition to chaos in clear detail. Libchaber knew that no clear image of the onset of turbulence had emerged in a laboratory, and he decided that his speck of a fluid cell would give a picture of the greatest possible clarity.

A NARROWING OF VISION helps keep science moving. By their lights, fluid dynamicists were correct to doubt the high level of precision that Swinney and Gollub claimed to have achieved in Couette flow. By their lights, mathematicians were correct to resent Ruelle, as they did. He had broken the rules. He had put forward an ambitious physical theory in the guise of a tight mathematical statement. He had made it hard to separate what he assumed from

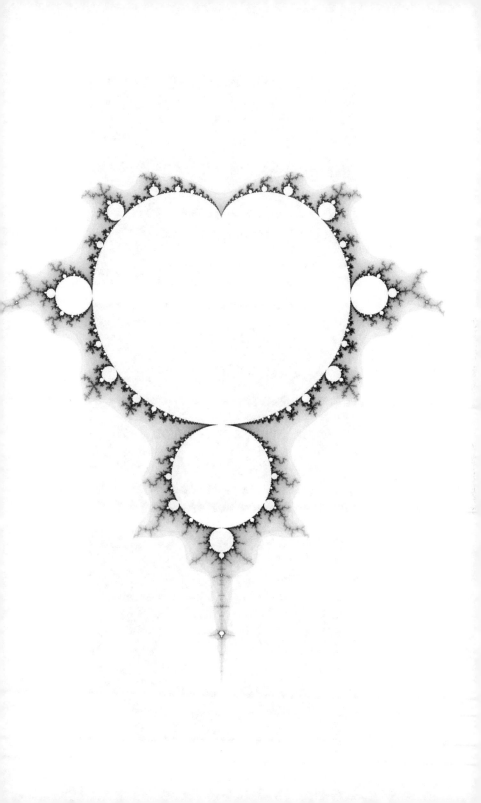

what he proved. The mathematician who refuses to endorse an idea until it meets the standard of theorem, proof, theorem, proof, plays a role that his discipline has written for him: consciously or not, he is standing watch against frauds and mystics. The journal editor who rejects new ideas because they are cast in an unfamiliar style may make his victims think that he is guarding turf on behalf of his established colleagues, but he, too, has a role to play in a community with reason to beware of the untried. "Science was constructed against a lot of nonsense," as Libchaber himself said. When his colleagues called Libchaber a mystic, the epithet was not always meant to be endearing.

He was an experimenter, careful and disciplined, known for precision in his prodding of matter. Yet he had a feeling for the abstract, ill-defined, ghostly thing called *flow*. Flow was shape plus change, motion plus form. A physicist, conceiving systems of differential equations, would call their mathematical movement a flow. Flow was a Platonic idea, assuming that change in systems reflected some reality independent of the particular instant. Libchaber embraced Plato's sense that hidden forms fill the universe. "But you know that they do! You have seen leaves. When you look at all the leaves, aren't you struck by the fact that the number of generic shapes is limited? You could easily draw the main shape. It would be of some interest to try to understand that. Or other shapes. In an experiment you have seen liquid penetrating into a liquid." His desk was strewn with pictures of such experiments, fat fractal fingers of liquid. "Now, in your kitchen, if you turn on your gas, you see that the flame is this shape again. It's very broad. It's universal. I don't care whether it's a burning flame or a liquid in a liquid or a solid growing crystal—what I'm interested in is this shape.

"There has been since the eighteenth century some kind of dream that science was missing the evolution of shape in space and the evolution of shape in time. If you think of a flow, you can think of a flow in many ways, flow in economics or a flow in history. First it may be laminar, then bifurcating to a more complicated state, perhaps with oscillations. Then it may be chaotic."

The universality of shapes, the similarities across scales, the recursive power of flows within flows—all sat just beyond reach of the standard differential-calculus approach to equations of change.

But that was not easy to see. Scientific problems are expressed in the available scientific language. So far, the twentieth century's best expression of Libchaber's intuition about flow needed the language of poetry. Wallace Stevens, for example, asserted a feeling about the world that stepped ahead of the knowledge available to physicists. He had an uncanny suspicion about flow, how it repeated itself while changing:

"The flecked river
Which kept flowing and never the same way twice, flowing
Through many places, as if it stood still in one."

Stevens's poetry often imparts a vision of tumult in atmosphere and water. It also conveys a faith about the invisible forms that order takes in nature, a belief

"that, in the shadowless atmosphere,
The knowledge of things lay round but unperceived."

When Libchaber and some other experimenters in the 1970s began looking into the motion of fluids, they did so with something approaching this subversive poetic intent. They suspected a connection between motion and universal form. They accumulated data in the only way possible, writing down numbers or recording them in a digital computer. But then they looked for ways to organize the data in ways that would reveal shapes. They hoped to express shapes in terms of motion. They were convinced that dynamical shapes like flames and organic shapes like leaves borrowed their form from some not-yet-understood weaving of forces. These experimenters, the ones who pursued chaos most relentlessly, succeeded by refusing to accept any reality that could be frozen motionless. Even Libchaber would not have gone so far as to express it in such terms, but their conception came close to what Stevens felt as an "insolid billowing of the solid":

"The vigor of glory, a glittering in the veins,
As things emerged and moved and were dissolved,

Either in distance, change or nothingness,
The visible transformations of summer night,

An argentine abstraction approaching form
And suddenly denying itself away."

FOR LIBCHABER, GOETHE, NOT STEVENS, supplied mystical inspiration. While Feigenbaum was looking through Harvard's library for Goethe's *Theory of Colors*, Libchaber had already managed to add to his collection an original edition of the even more obscure monograph *On the Transformation of Plants*. This was Goethe's sidelong assault on physicists who, he believed, worried exclusively about static phenomena rather than the vital forces and flows that produce the shapes we see from instant to instant. Part of Goethe's legacy—a negligible part, as far as literary historians were concerned—was a pseudoscientific following in Germany and Switzerland, kept alive by such philosophers as Rudolf Steiner and Theodor Schwenk. These men, too, Libchaber admired as much as a physicist could.

"Sensitive chaos"—*Das sensible Chaos*—was Schwenk's phrase for the relation between force and form. He used it for the title of a strange little book first published in 1965 and falling sporadically in and out of print thereafter. It was a book first about water. The English edition carried an admiring preface from Commandant Jacques Y. Cousteau and testimonials from the *Water Resources Bulletin* and the *Journal of the Institute of Water Engineers*. Little pretense at science marred Schwenk's exposition, and none at mathematics. Yet he observed flawlessly. He laid out a multitude of natural flowing shapes with an artist's eye. He assembled photographs and made dozens of precise drawings, like the sketches of a cell biologist peering through his first microscope. He had an open-mindedness and a naïveté that would have made Goethe proud.

Flow fills his pages. Great rivers like the Mississippi and the Bassin d'Arcachon in France meander in wide curves to the sea. In the sea itself, the Gulf Stream, too, meanders, making loops that swing east and west. It is a giant river of warm water amid cold, as Schwenk said, a river that "builds its own banks out of the cold water itself." When the flow itself is past or invisible, the evidence of flow remains. Rivers of air leave their mark on the desert sand, showing the waves. The flow of the ebbing tide inscribes a network of veins on a beach. Schwenk did not believe in coincidence. He believed in universal principles, and, more than universality, he believed in a certain spirit in nature that made his prose uncomfortably anthropomorphic. His "archetypal principle" was this: that flow "wants to realize itself, regardless of the surrounding material."

Within currents, he knew, there are secondary currents. Water moving down a meandering river flows, secondarily, around the river's axis, toward one bank, down to the riverbed, across toward the other bank, up toward the surface, like a particle spiraling around a doughnut. The trail of any water particle forms a string twisting around other strings. Schwenk had a topologist's imagination for such patterns. "This picture of strands twisted together in a spiral is only accurate with respect to the actual movement. One does often speak of 'strands' of water; they are however not really single strands but whole surfaces, interweaving spatially and flowing past each other." He saw rhythms competing in waves, waves overtaking one another, dividing surfaces, and boundary layers. He saw eddies and vortices and vortex trains, understanding them as the "rolling" of one surface about another. Here he came as close as a philosopher could to the physicist's conception of the dynamics of approaching turbulence. His artistic conviction assumed universality. To Schwenk, vortices meant instability, and instability meant that a flow was fighting an inequality within itself, and the inequality was "archetypal." The rolling of eddies, the unfurling of ferns, the creasing of mountain ranges, the hollowing of animal organs all followed one path, as he saw it. It had nothing to do with any particular medium, or any particular kind of difference. The inequalities could be slow and fast, warm and cold, dense and tenuous, salt and fresh, viscous and fluid, acid and alkaline. At the boundary, life blossoms.

Life, though, was D'Arcy Wentworth Thompson's territory. This extraordinary naturalist wrote in 1917: "It may be that all the laws of energy, and all the properties of matter, and all the chemistry of all the colloids are as powerless to explain the body as they are impotent to comprehend the soul. For my part, I think not." D'Arcy Thompson brought to the study of life exactly what Schwenk, fatally, lacked: mathematics. Schwenk argued by analogy. His case—spiritual, flowering, encyclopedic—finally came down to a display of similarities. D'Arcy Thompson's masterwork, On Growth and Form, shared something of Schwenk's mood and something of his method. The modern reader wonders how much to credit the meticulous pictures of multipronged falling droplets of liquid, hanging in sinuous tendrils, displayed next to astonishingly similar living jellyfish. Is this just a highbrow case of coincidence? If two forms look alike, must we look for like causes?

D'Arcy Thompson surely stands as the most influential biologist ever left on the fringes of legitimate science. The twentieth century's revolution in biology, well under way in his lifetime, passed him by utterly. He ignored chemistry, misunderstood the cell, and could not have predicted the explosive development of genetics. His writing, even in his time, seemed too classical and literary—too beautiful—to be reliably scientific. No modern bi-

Theodor Schwenk

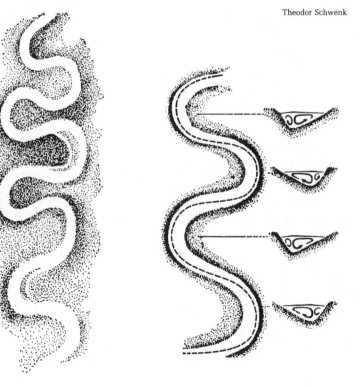

MEANDERING AND SPIRALING FLOWS. Theodor Schwenk depicted the currents of natural flows as strands with complicated secondary motions. "They are however not really single strands," he wrote, "but whole surfaces, interweaving spatially. . . ."

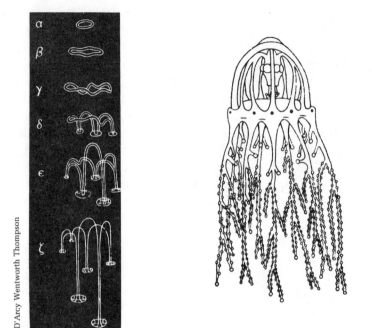

DROPS DESCENDING. D'Arcy Wentworth Thompson showed the hanging threads and columns made by ink drops falling through water (*left*) and by jellyfish (right). "An extremely curious result . . . is to show how sensitive these . . . drops are to physical conditions. For using the same gelatine all the while, and merely varying the density of the fluid in the third decimal place, we obtain a whole range of configurations, from the ordinary hanging drop to the same with a ribbed pattern. . . ."

ologist has to read D'Arcy Thompson. Yet somehow the greatest biologists find themselves drawn to his book. Sir Peter Medawar called it "beyond comparison the finest work of literature in all the annals of science that have been recorded in the English tongue." Stephen Jay Gould found no place better to turn for the intellectual pedigree of his own growing sense that nature constrains the shapes of things. Apart from D'Arcy Thompson, not many modern biologists had pursued the undeniable unity of living organisms. "Few had asked whether all the patterns might be reduced to a single system of generating forces," as Gould put it. "And few seemed to sense what significance such a proof of unity might possess for the science of organic form."

This classicist, polyglot, mathematician, zoologist tried to see life whole, just as biology was turning so productively toward methods that reduced organisms to their constituent functioning parts. Reductionism triumphed, most thrillingly in molecular biology but everywhere else as well, from evolution to medicine. How else to understand cells but by understanding membranes and nuclei and ultimately proteins, enzymes, chromosomes, and base pairs? When biology finally broached the interior workings of sinuses, retinas, nerves, brain tissue, it became unamusingly quaint to care about the *shape* of the skull. D'Arcy Thompson was the last to do so. He was also the last great biologist for many years to devote rhetorical energy to a careful discussion of *cause*, particularly the distinction between final cause and efficient or physical cause. Final cause is cause based on purpose or design: a wheel is round because that shape makes transportation possible. Physical cause is mechanical: the earth is round because gravity pulls a spinning fluid into a spheroid. The distinction is not always so obvious. A drinking glass is round because that is the most comfortable shape to hold or drink from. A drinking glass is round because that is the shape naturally assumed by spun pottery or blown glass.

In science, on the whole, physical cause dominates. Indeed, as astronomy and physics emerged from the shadow of religion, no small part of the pain came from discarding arguments by design, forward-looking teleology—the earth is what it is so that humanity can do what it does. In biology, however, Darwin firmly established teleology as the central mode of thinking about cause. The biological world may not fulfill God's design, but it fulfills a design shaped by natural selection. Natural selection operates not on genes or embryos, but on the final product. So an adaptationist explanation for the shape of an organism or the function of an organ always looks to its *cause*, not its physical cause but its final cause. Final cause survives in science wherever Darwinian thinking has become habitual. A modern anthropologist speculating about cannibalism or ritual sacrifice tends, rightly or wrongly, to ask only what purpose it serves. D'Arcy Thompson saw this coming. He begged that biology remember physical cause as well, mechanism and teleology together. He devoted himself to explaining the mathematical and physical forces that work on life. As adaptationism took hold, such explanations came to seem irrelevant. It became a rich and fruitful problem to explain a leaf in terms of how natural selection shaped such an effective solar panel. Only much later did some scientists start to puzzle again over the side of nature left unexplained. Leaves come in just a few shapes, of all the shapes imaginable; and the shape of a leaf is not dictated by its function.

The mathematics available to D'Arcy Thompson could not prove what he wanted to prove. The best he could do was draw, for example, skulls of related species with a crosshatching of coordinates, demonstrating that a simple geometric transformation turned one into the other. For simple organisms—with shapes so tantalizingly reminiscent of liquid jets, droplet splashes, and other manifestations of flow—he suspected physical causes, such as gravity and surface tension, that just could not do the formative work he asked of them. Why then, was Albert Libchaber thinking about *On Growth and Form* when he began his fluid experiments?

D'Arcy Thompson's intuition about the forces that shape life came closer than anything in the mainstream of biology to the perspective of dynamical systems. He thought of life as *life*, always in motion, always responding to rhythms—the "deep-seated rhythms of growth" which he believed created universal forms. He considered his proper study not just the material forms of things but their dynamics—"the interpretation, in terms of force, of the operations of Energy." He was enough of a mathematician to know that cataloguing shapes proved nothing. But he was enough of a poet to trust that neither accident nor purpose could explain the striking universality of forms he had assembled in his long years of gazing at nature. Physical laws must explain it, governing force and growth in ways that were just out of understanding's reach. Plato again. Behind the particular, visible shapes of matter must lie ghostly forms serving as invisible templates. Forms in motion.

LIBCHABER CHOSE LIQUID HELIUM for his experiment. Liquid helium has exceedingly low viscosity, so it will roll at the slightest push. The equivalent experiment in a medium-viscosity fluid like water or air would have taken a much larger box. With low viscosity, Libchaber made his experiment that much more sensitive to heating. To cause convection in his millimeter-wide cell, he had only to create a temperature difference of a thousandth of a degree between the top and bottom surfaces. That was why the cell had to be so tiny. In a larger box, where the liquid helium would have more room to roll, the equivalent motion would require even less heating, much less. In a box ten times larger in each direction, the size of a grape—a thousand times greater in volume—convection would begin with a heat differential of a millionth of a degree. Such minute temperature variations could not be controlled.

In the planning, in the design, in the construction, Libchaber and his engineer devoted themselves to eliminating any hint of messiness. In fact, they did all they could to eliminate the motion they were trying to study. Fluid motion, from smooth flow to turbulence, is thought of as motion through space. Its complexity appears as a spatial complexity, its disturbances and vortices as a spatial chaos. But Libchaber was looking for rhythms that would expose themselves as change over time. Time was the playing field and the yardstick. He squeezed space down nearly to a one-dimensional point. He was bringing to an extreme a technique that his predecessors in fluid experimentation had used, too. Everyone knew that an enclosed flow—Rayleigh-Bénard convection in a box or Couette-Taylor rotation in a cylinder—behaved measurably better than an open flow, like waves in the ocean or the air. In open flow, the boundary surface remains free, and the complexity multiplies.

Since convection in a rectilinear box produces rolls of fluid like hot dogs—or in this case like sesame seeds—he chose the dimensions of his cell carefully to allow precisely enough room for two rolls. The liquid helium would rise in the center, turn up and over to the left and right, and then descend on the outside edges of the cell. It was an arrested geometry. The wobbling would be confined. Clean lines and careful proportions would eliminate any extraneous fluctuations. Libchaber froze the space so that he could play with the time.

Once the experiment began, the helium rolling inside the cell inside the vacuum container inside the nitrogen bath, Libchaber would need some way to see what was happening. He embedded two microscopic temperature probes in the sapphire upper surface of the cell. Their output was recorded continuously by a pen plotter. Thus he could monitor the temperatures at two spots at the top of the fluid. It was so sensitive, so clever, another physicist said, that Libchaber succeeded in cheating nature.

This miniature masterpiece of precision took two years to explore fully, but it was, as he said, the right brush for his painting, not too grand or sophisticated. He finally saw everything. Running his experiment hour after hour, night and day, Libchaber found a more intricate pattern of behavior in the onset of turbulence than he had ever imagined. The full period-doubling cascade appeared. Libchaber confined and purified the motion of a fluid that rises when heated. The process begins with the first bifurcation, the onset of motion as soon as the bottom plate of high-purity copper heats up enough to overcome the tendency of the fluid to remain still. At a few degrees above absolute zero, a mere one-thousandth of a degree is enough. The liquid at the bottom warms and expands enough to become lighter than the cool liquid above. To let the warm liquid rise, the cool liquid must sink. Immediately, to let both motions occur, the liquid organizes itself into a pair of rolling cylinders. The rolls reach a constant speed, and the system settles into an equilibrium—a moving equilibrium, with heat energy being converted steadily into motion and dissipating through friction back to heat and passing out through the cool top plate.

So far, Libchaber was reproducing a well-known experiment in fluid mechanics, so well known that it was disdained. "It was classical physics," he said, "which unfortunately meant it was old, which meant it was uninteresting." It also happened to be precisely the flow that Lorenz had modeled with his system of three equations. But a real-world experiment—real liquid, a box cut by a machinist, a laboratory subject to the vibrations of Parisian traffic—already made the task of collecting data far more troublesome than simply generating numbers by a computer.

Experimenters like Libchaber used a simple pen plotter to record the temperature, as measured by a probe embedded in the top surface. In the equilibrium motion after the first bifurcation, the temperature at any one point remains steady, more or less, and the pen records a straight line. With more heating, more instability sets in. A kink develops in each roll, and the kink moves steadily back and forth. This wobble shows up as a changing temperature, up and down between two values. The pen now draws a wavy line across the paper.

From a simple temperature line, changing continuously and shaken by experimental noise, it becomes impossible to read the exact timing of new bifurcations or to deduce their nature. The line makes erratic peaks and valleys that seem almost as random as a stock market fever line. Libchaber analyzed such data by turning it into a spectrum diagram, meant to reveal the main frequencies hidden in the changing temperatures. Making a spectrum diagram of data from an experiment is like graphing the sound frequencies that make up a complex chord in a symphony. An uneven line of fuzziness always runs across the bottom of the graph—experimental noise. The main tones show up as vertical spikes: the louder the tone, the higher the spike. Similarly, if the data produce a dominant frequency—a rhythm peaking once a second, for example—then that frequency will show up as a spike on a spectrum diagram.

In Libchaber's experiment, as it happened, the first wavelength to appear was about two seconds. The next bifurcation brought a subtle change. The roll continued to wobble and the bolometer temperature continued to rise and fall with a dominant rhythm. But on odd cycles the temperature started going a bit higher than before, and on even cycles a bit lower. In fact, the maximum temperature split in two, so that there were two different maximums and two minimums. The pen line, though hard to read, developed a wobble on top of a wobble—a metawobble. On the spectrum diagram, that showed up more clearly. The old frequency was still strongly present, since the temperature still rose every two seconds. Now, however, a new frequency appeared at exactly half the old frequency, because the system had developed a component that repeated every four seconds. As the bifurcations continued, it was possible to distinguish a strangely consistent pattern: new frequencies appeared at double the old, so that the diagram filled in the quarters and the eighths and the sixteenths, starting to resemble a picket fence with alternating short and tall pickets.

Even to a man looking for hidden forms in messy data, tens and then hundreds of runs were necessary before the habits of

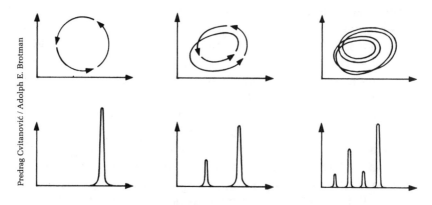

Predrag Cvitanović / Adolph E. Brotman

TWO WAYS OF SEEING A BIFURCATION. When an experiment like Libchaber's convection cell produces a steady oscillation, its phase-space portrait is a loop, repeating itself at regular intervals (*top left*). An experimenter
measuring the frequencies in the data will see a spectrum diagram with
a strong spike for this single rhythm. After a period-doubling bifurcation,
the system loops twice before repeating itself exactly (*center*), and now
the experimenter sees a new rhythm at half the frequency—twice the
period—of the original. New period-doublings fill in the spectrum diagram with more spikes.

this tiny cell started to come clear. Peculiar things could always
happen as Libchaber and his engineer slowly turned up the temperature and the system settled from one equilibrium into another.
Sometimes transient frequencies would appear, slide slowly across
the spectrum diagram, and disappear. Sometimes, the clean geometry notwithstanding, three rolls would develop instead of two—
and how could they know, really, what was happening inside that
tiny cell?

IF LIBCHABER HAD KNOWN then of Feigenbaum's discovery of
universality, he would have known exactly where to look for his
bifurcations and what to call them. By 1979 a growing group of
mathematicians and mathematically inclined physicists were pay-
ing attention to Feigenbaum's new theory. But the mass of scientists familiar with the problems of real physical systems believed
that they had good reason to withhold judgment. Complexity was

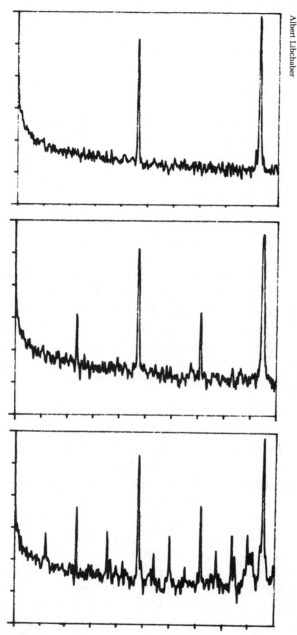

REAL-WORLD DATA CONFIRMING THEORY. Libchaber's spectrum diagrams showed vividly the precise pattern of period-doubling predicted by theory. The spikes of new frequencies stand out clearly above the experimental noise. Feigenbaum's scaling theory predicted not only when and where the new frequencies would arrive but also how strong they would be—their amplitudes.

one thing in the one-dimensional systems, the maps of May and Feigenbaum. It was surely something else in the two- or three- or four-dimensional systems of mechanical devices that an engineer could build. Those required serious differential equations, not just simple difference equations. And another chasm seemed to divide those low-dimensional systems from systems of fluid flow, which physicists thought of as potentially infinite-dimensional systems. Even a cell like Libchaber's, so carefully structured, had a virtual infinitude of fluid particles. Each particle represented at least the potential for independent motion. In some circumstances, any particle might be the locus of some new twist or vortex.

"The notion that the actual relevant meat-and-potatoes motion in such a system boils down to maps—nobody understood that," said Pierre Hohenberg of AT&T Bell Laboratories in New Jersey. Hohenberg became one of the very few physicists to follow the new theory and the new experiments together. "Feigenbaum may have dreamt of that, but he certainly didn't say it. Feigenbaum's work was about maps. Why should physicists be interested in maps?—it's a game. Really, as long as they were playing around with maps, it seemed pretty remote from what we wanted to understand.

"But when it was seen in experiments, that's when it really became exciting. The miracle is that, in systems that are interesting, you can still understand behavior in detail by a model with a small number of degrees of freedom."

It was Hohenberg, in the end, who brought the theorist and the experimenter together. He ran a workshop at Aspen in the summer of 1979, and Libchaber was there. (Four years earlier, at the same summer workshop, Feigenbaum had listened to Steve Smale talk about a number—just a number—that seemed to pop up when a mathematician looked at the transition to chaos in a certain equation.) When Libchaber described his experiments with liquid helium, Hohenberg took note. On his way home, Hohenberg happened to stop and see Feigenbaum in New Mexico. Not long after, Feigenbaum paid a call on Libchaber in Paris. They stood amid the scattered parts and instruments of Libchaber's laboratory. Libchaber proudly displayed his tiny cell and let Feigenbaum explain his latest theory. Then they walked through the Paris streets looking for the best possible cup of coffee. Libchaber remembered later how surprised he was to see a theorist so young and so, he would say, *lively.*

THE LEAP FROM MAPS TO FLUID FLOW seemed so great that even those most responsible sometimes felt it was like a dream. How nature could tie such complexity to such simplicity was far from obvious. "You have to regard it as a kind of miracle, not like the usual connection between theory and experiment," Jerry Gollub said. Within a few years, the miracle was being repeated again and again in a vast bestiary of laboratory systems: bigger fluid cells with water and mercury, electronic oscillators, lasers, even chemical reactions. Theorists adapted Feigenbaum's techniques and found other mathematical routes to chaos, cousins of period-doubling: such patterns as intermittency and quasiperiodicity. These, too, proved universal in theory and experiment.

The experimenters' discoveries helped set in motion the era of computer experimentation. Physicists discovered that computers produced the same qualitative pictures as real experiments, and produced them millions of times faster and more reliably. To many, even more convincing than Libchaber's results was a fluid model created by Valter Franceschini of the University of Modena, Italy—a system of five differential equations that produced attractors and period-doubling. Franceschini knew nothing of Feigenbaum, but his complex, many-dimensional model produced the same constants Feigenbaum had found in one-dimensional maps. In 1980 a European group provided a convincing mathematical explanation: dissipation bleeds a complex system of many conflicting motions, eventually bringing the behavior of many dimensions down to one.

Outside of computers, to find a strange attractor in a fluid experiment remained a serious challenge. It occupied experimenters like Harry Swinney well into the 1980s. And when the experimenters finally succeeded, the new computer experts often belittled their results as just the rough, predictable echoes of the magnificently detailed pictures their graphics terminals were already churning out. In a computer experiment, when you generated your thousands or millions of data points, patterns made themselves more or less apparent. In a laboratory, as in the real world, useful information had to be distinguished from noise. In a computer experiment data flowed like wine from a magic chalice. In a laboratory experiment you had to fight for every drop.

Still, the new theories of Feigenbaum and others would not have captured so wide a community of scientists on the strength of computer experiments alone. The modifications, the compromises, the approximations needed to digitize systems of nonlinear differential equations were too suspect. Simulations break reality into chunks, as many as possible but always too few. A computer model is just a set of arbitrary rules, chosen by programmers. A real-world fluid, even in a stripped-down millimeter cell, has the undeniable potential for all the free, untrammeled motion of natural disorder. It has the potential for surprise.

In the age of computer simulation, when flows in everything from jet turbines to heart valves are modeled on supercomputers, it is hard to remember how easily nature can confound an experimenter. In fact, no computer today can completely simulate even so simple a system as Libchaber's liquid helium cell. Whenever a good physicist examines a simulation, he must wonder what bit of reality was left out, what potential surprise was sidestepped. Libchaber liked to say that he would not want to fly in a simulated airplane—he would wonder what had been missed. Furthermore, he would say that computer simulations help to build intuition or to refine calculations, but they do not give birth to genuine discovery. This, at any rate, is the experimenter's creed.

His experiment was so immaculate, his scientific goals so abstract, that there were still physicists who considered Libchaber's work more philosophy or mathematics than physics. He believed, in turn, that the ruling standards of his field were reductionist, giving primacy to the properties of atoms. "A physicist would ask me, How does this atom come here and stick there? And what is the sensitivity to the surface? And can you write the Hamiltonian of the system?

"And if I tell him, I don't care, what interests me is this *shape*, the mathematics of the shape and the evolution, the bifurcation from this shape to that shape to this shape, he will tell me, that's not physics, you are doing mathematics. Even today he will tell me that. Then what can I say? Yes, of course, I am doing mathematics. But it is relevant to what is around us. That is nature, too."

The patterns he found were indeed abstract. They were mathematical. They said nothing about the properties of liquid helium or copper or about the behavior of atoms near absolute zero. But they were the patterns that Libchaber's mystical forebears had dreamed of. They made legitimate a realm of experimentation in which many scientists, from chemists to electrical engineers, soon became explorers, seeking out the new elements of motion. The patterns were there to see the first time he succeeded in raising the temperature enough to isolate the first period-doubling, and the next, and the next. According to the new theory, the bifurcations should have produced a geometry with precise scaling, and that was just what Libchaber saw, the universal Feigenbaum constants turning in that instant from a mathematical ideal to a physical reality, measurable and reproducible. He remembered the feeling long afterward, the eerie witnessing of one bifurcation after another and then the realization that he was seeing an infinite cascade, rich with structure. It was, as he said, amusing.

Images of Chaos

What else, when chaos draws all forces inward
To shape a single leaf.

—CONRAD AIKEN

MICHAEL BARNSLEY MET Mitchell Feigenbaum at a conference in Corsica in 1979. That was when Barnsley, an Oxford-educated mathematician, learned about universality and period-doubling and infinite cascades of bifurcations. A good idea, he thought, just the sort of idea that was sure to send scientists rushing to cut off pieces for themselves. For his part, Barnsley thought he saw a piece that no one else had noticed.

Where were these cycles of 2, 4, 8, 16, these Feigenbaum sequences, coming from? Did they appear by magic out of some mathematical void, or did they suggest the shadow of something deeper still. Barnsley's intuition was that they must be part of some fabulous fractal object so far hidden from view.

For this idea, he had a context, the numerical territory known as the complex plane. In the complex plane, the numbers from minus infinity to infinity—all the real numbers, that is—lie on a line stretching from the far west to the far east, with zero at the center. But this line is only the equator of a world that also stretches to infinity in the north and the south. Each number is composed of two parts, a *real* part, corresponding to east-west longitude, and an *imaginary* part, corresponding to north-south latitude. By convention, these complex numbers are written this way: $2 + 3i$, the i signifying the imaginary part. The two parts give each number a unique address in this two-dimensional plane. The original line of real numbers, then, is just a special case, the set of numbers whose imaginary part equals zero. In the complex plane, to look only at the real numbers—only at points on the equator—would be to limit one's vision to occasional intersections of shapes that might reveal other secrets when viewed in two dimensions. So Barnsley suspected.

The names *real* and *imaginary* originated when ordinary numbers did seem more real than this new hybrid, but by now the names were recognized as quite arbitrary, both sorts of numbers being just as real and just as imaginary as any other sort. Historically, imaginary numbers were invented to fill the conceptual vacuum produced by the question: What is the square root of a negative number? By convention, the square root of -1 is i, the square root of -4 is $2i$, and so on. It was only a short step to the realization that combinations of real and imaginary numbers allowed new kinds of calculations with polynomial equations. Complex numbers can be added, multiplied, averaged, factored, integrated. Just about any calculation on real numbers can be tried on complex numbers as well. Barnsley, when he began translating Feigenbaum functions into the complex plane, saw outlines emerging of a fantastical family of shapes, seemingly related to the dynamical ideas intriguing experimental physicists, but also startling as mathematical constructs.

These cycles do not appear out of thin air after all, he realized. They fall into the real line off the complex plane, where, if you look, there is a constellation of cycles, of all orders. There always *was* a two-cycle, a three-cycle, a four-cycle, floating just out of sight until they arrived on the real line. Barnsley hurried back from Corsica to his office at the Georgia Institute of Technology and produced a paper. He shipped it off to *Communications in Mathematical Physics* for publication. The editor, as it happened, was David Ruelle, and Ruelle had some bad news. Barnsley had unwittingly rediscovered a buried fifty-year-old piece of work by a French mathematician. "Ruelle shunted it back to me like a hot potato and said, 'Michael, you're talking about Julia sets,' " Barnsley recalled.

Ruelle added one piece of advice: "Get in touch with Mandelbrot."

JOHN HUBBARD, AN AMERICAN MATHEMATICIAN with a taste
for fashionable bold shirts, had been teaching elementary calculus
to first-year university students in Orsay, France, three years before. Among the standard topics that he covered was Newton's
method, the classic scheme for solving equations by making successively better approximations. Hubbard was a little bored with
standard topics, however, and for once he decided to teach Newton's method in a way that would force his students to think.

Newton's method is old, and it was already old when Newton
invented it. The ancient Greeks used a version of it to find square
roots. The method begins with a guess. The guess leads to a better
guess, and the process of iteration zooms in on an answer like a
dynamical system seeking its steady state. The process is fast, the
number of accurate decimal digits generally doubling with each
step. Nowadays, of course, square roots succumb to more analytic
methods, as do all roots of degree-two polynomial equations—
those in which variables are raised only to the second power. But
Newton's method works for higher-degree polynomial equations
that cannot be solved directly. The method also works beautifully
in a variety of computer algorithms, iteration being, as always, the
computer's forte. One tiny awkwardness about Newton's method
is that equations usually have more than one solution, particularly
when complex solutions are included. *Which* solution the method
finds depends on the initial guess. In practical terms, students
find that this is no problem at all. You generally have a good idea
of where to start, and if your guess seems to be converging to the
wrong solution, you just start someplace else.

One might ask exactly what sort of route Newton's method
traces as it winds toward a root of a degree-two polynomial on
the complex plane. One might answer, thinking geometrically, that
the method simply seeks out whichever of the two roots is closer
to the initial guess. That is what Hubbard told his students at
Orsay when the question arose one day.

"Now, for equations of, say, degree three, the situation seems
more complicated," Hubbard said confidently. "I will think of it
and tell you next week."

He still presumed that the hard thing would be to teach his
students how to calculate the iteration and that making the initial

guess would be easy. But the more he thought about it, the less he knew—about what constituted an intelligent guess or, for that matter, about what Newton's method really did. The obvious geometric guess would be to divide the plane into three equal pie wedges, with one root inside each wedge, but Hubbard discovered that that would not work. Strange things happened near the boundaries. Furthermore, Hubbard discovered that he was not the first mathematician to stumble on this surprisingly difficult question. Arthur Cayley had tried in 1879 to move from the manageable second-degree case to the frighteningly intractable third-degree case. But Hubbard, a century later, had a tool at hand that Cayley lacked.

Hubbard was the kind of rigorous mathematician who despised guesses, approximations, half-truths based on intuition rather than proof. He was the kind of mathematician who would continue to insist, twenty years after Edward Lorenz's attractor entered the literature, that no one really *knew* whether those equations gave rise to a strange attractor. It was unproved conjecture. The familiar double spiral, he said, was not proof but mere evidence, something computers drew.

Now, in spite of himself, Hubbard began using a computer to do what the orthodox techniques had not done. The computer would prove nothing. But at least it might unveil the truth so that a mathematician could know what it was he should try to prove. So Hubbard began to experiment. He treated Newton's method not as a way of solving problems but as a problem in itself. Hubbard considered the simplest example of a degree-three polynomial, the equation $x^3 - 1 = 0$. That is, find the cube root of 1. In real numbers, of course, there is just the trivial solution: 1. But the polynomial also has two complex solutions: $-1/2 + i\sqrt{3}/2$, and $-1/2 - i\sqrt{3}/2$. Plotted in the complex plane, these three roots mark an equilateral triangle, with one point at three o'clock, one at seven o'clock, and one at eleven o'clock. Given any complex number as a starting point, the question was to see *which* of the three solutions Newton's method would lead to. It was as if Newton's method were a dynamical system and the three solutions were three attractors. Or it was as if the complex plane were a smooth surface sloping down toward three deep valleys. A marble starting from anywhere on the plane should roll into one of the valleys—but which? Hubbard set about sampling the infinitude of points that make up the plane. He had his computer sweep from point to point, calculating the flow of Newton's method for each one, and color-coding the results. Starting points that led to one solution were all colored blue. Points that led to the second solution were red, and points that led to the third were green. In the crudest approximation, he found, the dynamics of Newton's method did indeed divide the plane into three pie wedges. Generally the points *near* a particular solution led quickly into that solution. But systematic computer exploration showed complicated underlying organization that could never have been seen by earlier mathematicians, able only to calculate a point here and a point there. While some starting guesses converged quickly to a root, others bounced around seemingly at random before finally converging to a solution. Sometimes it seemed that a point could fall into a cycle that would repeat itself forever—a periodic cycle—without ever reaching one of the three solutions.

As Hubbard pushed his computer to explore the space in finer and finer detail, he and his students were bewildered by the picture that began to emerge. Instead of a neat ridge between the blue and red valleys, for example, he saw blotches of green, strung together like jewels. It was as if a marble, caught between the conflicting tugs of two nearby valleys, would end up in the third and most distant valley instead. A boundary between two colors never quite forms. On even closer inspection, the line between a green blotch and the blue valley proved to have patches of red. And so on—the boundary finally revealed to Hubbard a peculiar property that would seem bewildering even to someone familiar with Mandelbrot's monstrous fractals: *no point serves as a boundary between just two colors.* Wherever two colors try to come together, the third always inserts itself, with a series of new, self-similar intrusions. Impossibly, every boundary point borders a region of each of the three colors.

Hubbard embarked on a study of these complicated shapes and their implications for mathematics. His work and the work of his colleagues soon became a new line of attack on the problem of dynamical systems. He realized that the mapping of Newton's

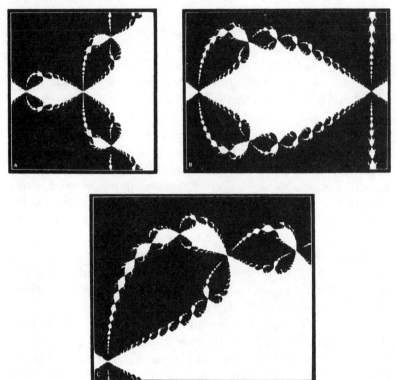

Heinz-Otto Peitgen, Peter H. Richter

BOUNDARIES OF INFINITE COMPLEXITY. When a pie is cut into three slices, they meet at a single point, and the boundaries between any two slices are simple. But many processes of abstract mathematics and real-world physics turn out to create boundaries that are almost unimaginably complex.

Above, Newton's method applied to finding the cube root of −1 divides the plane into three identical regions, one of which is shown in white. All white points are "attracted" to the root lying in the largest white area; all black points are attracted to one of the other two roots. The boundary has the peculiar property that every point on it borders all three regions. And, as the insets show, magnified segments reveal a fractal structure, repeating the basic pattern on smaller and smaller scales.

method was just one of a whole unexplored family of pictures that reflected the behavior of forces in the real world. Michael Barnsley was looking at other members of the family. Benoit Mandelbrot, as both men soon learned, was discovering the granddaddy of all these shapes.

THE MANDELBROT SET IS the most complex object in mathematics, its admirers like to say. An eternity would not be enough time to see it all, its disks studded with prickly thorns, its spirals and filaments curling outward and around, bearing bulbous molecules that hang, infinitely variegated, like grapes on God's personal vine. Examined in color through the adjustable window of a computer screen, the Mandelbrot set seems more fractal than fractals, so rich is its complication across scales. A cataloguing of the different images within it or a numerical description of the set's outline would require an infinity of information. But here is a paradox: to send a full description of the set over a transmission line requires just a few dozen characters of code. A terse computer program contains enough information to reproduce the entire set. Those who were first to understand the way the set commingles complexity and simplicity were caught unprepared—even Mandelbrot. The Mandelbrot set became a kind of public emblem for chaos, appearing on the glossy covers of conference brochures and engineering quarterlies, forming the centerpiece of an exhibit of computer art that traveled internationally in 1985 and 1986. Its beauty was easy to feel from these pictures; harder to grasp was the meaning it had for the mathematicians who slowly understood it.

Many fractal shapes can be formed by iterated processes in the complex plane, but there is just one Mandelbrot set. It started appearing, vague and spectral, when Mandelbrot tried to find a way of generalizing about a class of shapes known as Julia sets. These were invented and studied during World War I by the French mathematicians Gaston Julia and Pierre Fatou, laboring without the pictures that a computer could provide. Mandelbrot had seen their modest drawings and read their work—already obscure—when he was twenty years old. Julia sets, in a variety of guises, were precisely the objects intriguing Barnsley. Some Julia sets are like circles that have been pinched and deformed in many places to give them a fractal structure. Others are broken into regions, and still others are disconnected dusts. But neither words nor the concepts of Euclidean geometry serve to describe them. The French mathematician Adrien Douady said: "You obtain an incredible

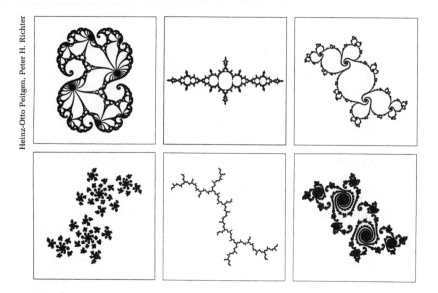

AN ASSORTMENT OF JULIA SETS.

variety of Julia sets: some are a fatty cloud, others are a skinny bush of brambles, some look like the sparks which float in the air after a firework has gone off. One has the shape of a rabbit, lots of them have sea-horse tails."

In 1979 Mandelbrot discovered that he could create one image in the complex plane that would serve as a catalogue of Julia sets, a guide to each and every one. He was exploring the iteration of complicated processes, equations with square roots and sines and cosines. Even after building his intellectual life around the proposition that simplicity breeds complexity, he did not immediately understand how extraordinary was the object hovering just beyond the view of his computer screens at IBM and Harvard. He pressed his programmers hard for more detail, and they sweated over the allocation of already strained memory, the new interpolation of points on an IBM mainframe computer with a crude black and white display tube. To make matters worse, the programmers always had to stand guard against a common pitfall of computer exploration, the production of "artifacts," features that sprang solely from some quirk of the machine and would disappear when a program was written differently.

Then Mandelbrot turned his attention to a simple mapping that was particularly easy to program. On a rough grid, with a program that repeated the feedback loop just a few times, the first outlines of disks appeared. A few lines of pencil calculation showed that the disks were mathematically real, not just products of some computational oddity. To the right and left of the main disks, hints of more shapes appeared. In his mind, he said later, he saw more: a hierarchy of shapes, atoms sprouting smaller atoms ad infinitum. And where the set intersected the real line, its successively smaller disks scaled with a geometric regularity that dynamicists now recognized: the Feigenbaum sequence of bifurcations.

That encouraged him to push the computation further, refining those first crude images, and he soon discovered dirt cluttering the edge of the disks and also floating in the space nearby. As he tried calculating in finer and finer detail, he suddenly felt that his string of good luck had broken. Instead of becoming sharper, the pictures became messier. He headed back to IBM's Westchester County research center to try computing power on a proprietary scale that Harvard could not match. To his surprise, the growing messiness was the sign of something real. Sprouts and tendrils spun languidly away from the main island. Mandelbrot saw a seemingly smooth boundary resolve itself into a chain of spirals like the tails of sea horses. The irrational fertilized the rational.

The Mandelbrot set is a collection of points. Every point in the complex plane—that is, every complex number—is either in the set or outside it. One way to define the set is in terms of a test for every point, involving some simple iterated arithmetic. To test a point, take the complex number; square it; add the original number; square the result; add the original number; square the result—and so on, over and over again. If the total runs away to infinity, then the point is not in the Mandelbrot set. If the total remains finite (it could be trapped in some repeating loop, or it could wander chaotically), then the point is in the Mandelbrot set.

This business of repeating a process indefinitely and asking whether the result is infinite resembles feedback processes in the everyday world. Imagine that you are setting up a microphone, amplifier, and speakers in an auditorium. You are worried about the squeal of sonic feedback. If the microphone picks up a loud

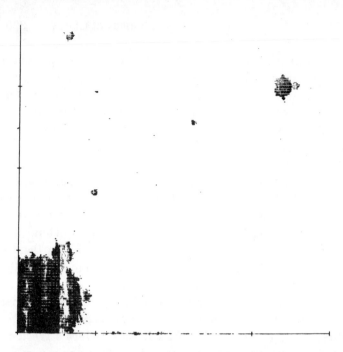

THE MANDELBROT SET EMERGES. In Benoit Mandelbrot's first crude computer printouts, a rough structure appeared, gaining more detail as the quality of the computation improved. Were the buglike, floating "molecules" isolated islands? Or were they attached to the main body by filaments too fine to be observed? It was impossible to tell.

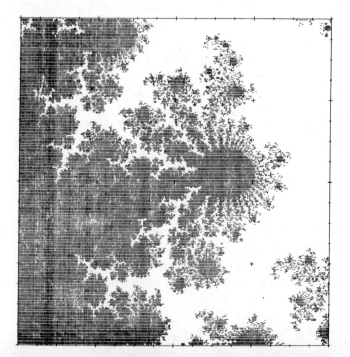

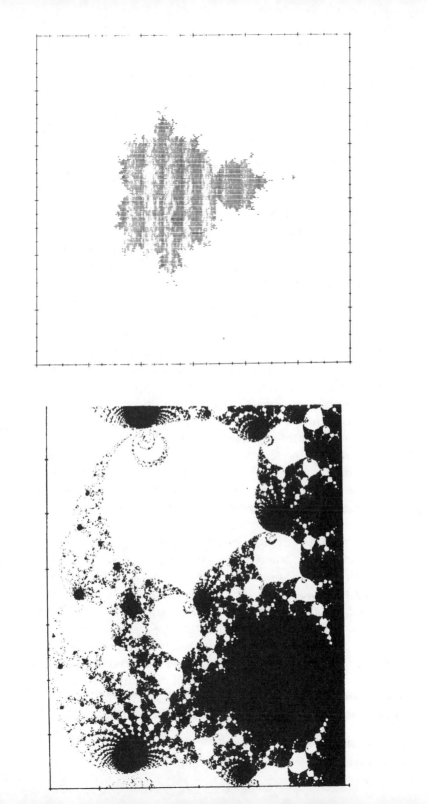

enough noise, the amplified sound from the speakers will feed back into the microphone in an endless, ever louder loop. On the other hand, if the sound is small enough, it will just die away to nothing. To model this feedback process with numbers, you might take a starting number, multiply it by itself, multiply the result by itself, and so on. You would discover that large numbers lead quickly to infinity: 10, 100, 10,000 . . . . But small numbers lead to zero: $\frac{1}{2}$, $\frac{1}{4}$, $\frac{1}{16}$ . . . . To make a geometric picture, you define a collection of all the points that, when fed into this equation, do not run away to infinity. Consider the points on a line from zero upward. If a point produces a squeal of feedback, color it white. Otherwise color it black. Soon enough, you will have a shape that consists of a black line from 0 to 1.

For a one-dimensional process, no one need actually resort to experimental trial. It is easy enough to establish that numbers greater than one lead to infinity and the rest do not. But in the two dimensions of the complex plane, to deduce a shape defined by an iterated process, knowing the equation is generally not enough. Unlike the traditional shapes of geometry, circles and ellipses and parabolas, the Mandelbrot set allows no shortcuts. The only way to see what kind of shape goes with a particular equation is by trial and error, and the trial-and-error style brought the explorers of this new terrain closer in spirit to Magellan than to Euclid.

Joining the world of shapes to the world of numbers in this way represented a break with the past. New geometries always begin when someone changes a fundamental rule. *Suppose space can be curved instead of flat*, a geometer says, and the result is a weird curved parody of Euclid that provides precisely the right framework for the general theory of relativity. Suppose space can have four dimensions, or five, or six. Suppose the number expressing *dimension* can be a fraction. Suppose shapes can be twisted, stretched, knotted. Or, now, suppose shapes are defined, not by solving an equation once, but by iterating it in a feedback loop.

Julia, Fatou, Hubbard, Barnsley, Mandelbrot—these mathematicians changed the rules about how to make geometrical shapes. The Euclidean and Cartesian methods of turning equations into curves are familiar to anyone who has studied high school geometry or found a point on a map using two coordinates. Standard geometry takes an equation and asks for the set of numbers that *satisfy* it. The solutions to an equation like $x^2 + y^2 = 1$, then, form a shape, in this case a circle. Other simple equations produce other pictures, the ellipses, parabolas, and hyperbolas of conic sections or even the more complicated shapes produced by differential equations in phase space. But when a geometer iterates an equation instead of solving it, the equation becomes a process instead of a description, dynamic instead of static. When a number goes into the equation, a new number comes out; the new number goes in, and so on, points hopping from place to place. A point is plotted not when it satisfies the equation but when it produces a certain kind of behavior. One behavior might be a steady state. Another might be a convergence to a periodic repetition of states. Another might be an out-of-control race to infinity.

Before computers, even Julia and Fatou, who understood the possibilities of this new kind of shape-making, lacked the means of making it a science. With computers, trial-and-error geometry became possible. Hubbard explored Newton's method by calculating the behavior of point after point, and Mandelbrot first viewed his set the same way, using a computer to sweep through the points of the plane, one after another. Not all the points, of course. Time and computers being finite, such calculations use a grid of points. A finer grid gives a sharper picture, at the expense of longer computation. For the Mandelbrot set, the calculation was simple, because the process itself was so simple: the iteration in the complex plane of the mapping $z \rightarrow z^2 + c$. Take a number, multiply it by itself, and add the original number.

As Hubbard grew comfortable with this new style of exploring shapes by computer, he also brought to bear an innovative mathematical style, applying the methods of complex analysis, an area of mathematics that had not been applied to dynamical systems before. Everything was coming together, he felt. Separate disciplines within mathematics were converging at a crossroads. He knew it would not suffice to *see* the Mandelbrot set; before he was done, he wanted to understand it, and indeed, he finally claimed that he did understand it.

If the boundary were merely fractal in the sense of Mandelbrot's turn-of-the-century monsters, then one picture would look more or less like the last. The principle of self-similarity at different scales would make it possible to predict what the electronic microscope would see at the next level of magnification. Instead, each foray deeper into the Mandelbrot set brought new surprises. Mandelbrot started worrying that he had offered too restrictive a definition of *fractal*; he certainly wanted the word to apply to this new object. The set did prove to contain, when magnified enough, rough copies of itself, tiny buglike objects floating off from the main body, but greater magnification showed that none of these molecules exactly matched any other. There were always new kinds of sea horses, new curling hothouse species. In fact, no part of the set exactly resembles any other part, at *any* magnification.

The discovery of floating molecules raised an immediate problem, though. Was the Mandelbrot set connected, one continent with far-flung peninsulas? Or was it a dust, a main body surrounded by fine islands? It was far from obvious. No guidance came from the experience with Julia sets, because Julia sets came in both flavors, some whole shapes and some dusts. The dusts, being fractal, have the peculiar property that no two pieces are "together"—because every piece is separated from every other by a region of empty space—yet no piece is "alone," since whenever you find one piece, you can always find a group of pieces arbitrarily close by. As Mandelbrot looked at his pictures, he realized that computer experimentation was failing to settle this fundamental question. He focused more sharply on the specks hovering about the main body. Some disappeared, but others grew into clear near-replicas. They seemed independent. But possibly they were connected by lines so thin that they continued to escape the lattice of computed points.

Douady and Hubbard used a brilliant chain of new mathematics to prove that every floating molecule does indeed hang on a filigree that binds it to all the rest, a delicate web springing from tiny outcroppings on the main set, a "devil's polymer," in Mandelbrot's phrase. The mathematicians proved that any segment—no matter where, and no matter how small—would, when blown up by the computer microscope, reveal new molecules, each resembling the main set and yet not quite the same. Every new molecule would be surrounded by its own spirals and flame-like projections, and those, inevitably, would reveal molecules tinier still, always similar, never identical, fulfilling some mandate of infinite variety, a miracle of miniaturization in which every new detail was sure to be a universe of its own, diverse and entire.

"EVERYTHING WAS VERY GEOMETRIC straight-line approaches," said Heinz-Otto Peitgen. He was talking about modern art. "The work of Josef Albers, for example, trying to discover the relation of colors, this was essentially just squares of different colors put onto each other. These things were very popular. If you look at it now it seems to have passed. People don't like it any more. In Germany they built huge apartment blocks in the Bauhaus style and people move out, they don't like to live there. There are very deep reasons, it seems to me, in society right now to dislike some aspects of our conception of nature." Peitgen had been helping a visitor select blowups of regions of the Mandelbrot set, Julia sets, and other complex iterative processes, all exquisitely colored. In his small California office he offered slides, large transparencies, even a Mandelbrot set calendar. "The deep enthusiasm we have has to do with this different perspective of looking at nature. What is the true aspect of the natural object? The tree, let's say—what is important? Is it the straight line, or is it the fractal object?" At Cornell, meanwhile, John Hubbard was struggling with the demands of commerce. Hundreds of letters were flowing into the mathematics department to request Mandelbrot set pictures, and he realized he had to create samples and price lists. Dozens of images were already calculated and stored in his computers, ready for instant display, with the help of the graduate students who remembered the technical detail. But the most spectacular pictures, with the finest resolution and the most vivid coloration, were coming from two Germans, Peitgen and Peter H. Richter, and their team of scientists at the University of Bremen, with the enthusiastic sponsorship of a local bank.

Peitgen and Richter, one a mathematician and the other a physicist, turned their careers over to the Mandelbrot set. It held a universe of ideas for them: a modern philosophy of art, a justification of the new role of experimentation in mathematics, a way of bringing complex systems before a large public. They published glossy catalogs and books, and they traveled around the world with a gallery exhibit of their computer images. Richter had come to complex systems from physics by way of chemistry and then biochemistry, studying oscillations in biological pathways. In a series of papers on such phenomena as the immune system and the conversion of sugar into energy by yeast, he found that oscillations often governed the dynamics of processes that were customarily viewed as static, for the good reason that living systems cannot easily be opened up for examination in real time. Richter kept clamped to his windowsill a well-oiled double pendulum, his "pet dynamical system," custom-made for him by his university machine shop. From time to time he would set it spinning in chaotic nonrhythms that he could emulate on a computer as well. The dependence on initial conditions was so sensitive that the gravitational pull of a single raindrop a mile away mixed up the motion within fifty or sixty revolutions, about two minutes. His multicolor graphic pictures of the phase space of this double pendulum showed the mingled regions of periodicity and chaos, and he used the same graphic techniques to display, for example, idealized regions of magnetization in a metal and also to explore the Mandelbrot set.

For his colleague Peitgen the study of complexity provided a chance to create new traditions in science instead of just solving problems. "In a brand new area like this one, you can start thinking today and if you are a good scientist you might be able to come up with interesting solutions in a few days or a week or a month," Peitgen said. The subject is unstructured.

"In a structured subject, it is known what is known, what is unknown, what people have already tried and doesn't lead anywhere. There you have to work on a problem which is known to be a problem, otherwise you get lost. But a problem which is known to be a problem must be hard, otherwise it would already have been solved."

Peitgen shared little of the mathematicians' unease with the use of computers to conduct experiments. Granted, every result must eventually be made rigorous by the standard methods of proof, or it would not be mathematics. To see an image on a graphics screen does not guarantee its existence in the language of theorem and proof. But the very availability of that image was enough to change the evolution of mathematics. Computer exploration was giving mathematicians the freedom to take a more natural path, Peitgen believed. Temporarily, for the moment, a mathematician could suspend the requirement of rigorous proof. He could go wherever experiments might lead him, just as a physicist could. The numerical power of computation and the visual cues to intuition would suggest promising avenues and spare the mathematician blind alleys. Then, new paths having been found and new objects isolated, a mathematician could return to standard proofs. "Rigor is the strength of mathematics," Peitgen said. "That we can continue a line of thought which is absolutely guaranteed—mathematicians never want to give that up. But you can look at situations that can be understood *partially* now and with rigor perhaps in future generations. Rigor, yes, but not to the extent that I drop something just because I can't do it *now*."

By the 1980s a home computer could handle arithmetic precise enough to make colorful pictures of the set, and hobbyists quickly found that exploring these pictures at ever-greater magnification gave a vivid sense of expanding scale. If the set were thought of as a planet-sized object, a personal computer could show the whole object, or features the size of cities, or the size of buildings, or the size of rooms, or the size of books, or the size of letters, or the size of bacteria, or the size of atoms. The people who looked at such pictures saw that all the scales had similar patterns, yet every scale was different. And all these microscopic landscapes were generated by the same few lines of computer code.*

---

*A Mandelbrot set program needs just a few essential pieces. The main engine is a loop of instructions that takes its starting complex number and applies the arithmetical rule to it. For the Mandelbrot set, the rule is this: $z \rightarrow z^2 + c$, where z begins at zero and c is the complex number corresponding to the point being tested. So, *take 0, multiply it by itself, and add the starting number; take the result—the starting number—multiply it by itself, and add the starting number; take the new result, multiply it by itself, and add the starting number.* Arithmetic with complex numbers is straightforward. A complex number is written with two parts: for example, $2 + 3i$ (the address for the point at 2 east and 3 north on the complex plane). To add a pair of complex numbers, you just add the real parts to get a new real part and the imaginary parts to get a new imaginary part:

$$\begin{array}{r} 2 + 4i \\ + \underline{9 - 2i} \\ 11 + 2i \end{array}$$

To multiply two complex numbers, you multiply each part of one number by each

THE BOUNDARY IS WHERE a Mandelbrot set program spends most of its time and makes all of its compromises. There, when 100 or 1,000 or 10,000 iterations fail to break away, a program still cannot be absolutely certain that a point falls inside the step. Who knows what the millionth iteration will bring? So the programs that made the most striking, most deeply magnified pictures of the set ran on heavy mainframe computers, or computers devoted to parallel processing, with thousands of individual brains performing the same arithmetic in lock step. The boundary is where points are slowest to escape the pull of the set. It is as if they are balanced between competing attractors, one at zero and the other, in effect, ringing the set at a distance of infinity.

When scientists moved from the Mandelbrot set itself to new problems of representing real physical phenomena, the qualities

---

part of the other and add the four results together. Because i multiplied by itself equals -1, by the original definition of imaginary numbers, one term of the result collapses into another.

$$
\begin{array}{r}
2 + 3i \\
\times\ 2 + 3i \\
\hline
6i + 9i^2 \\
4 + 6i \\
\hline
4 + 12i + 9i^2 \\
= 4 + 12i - 9 \\
= -5 + 12i
\end{array}
$$

To break out of this loop, the program needs to watch the running total. If the total heads off to infinity, moving farther and farther from the center of the plane, the original point does not belong to the set, and if the running total becomes greater than 2 or smaller than -2 in either its real or imaginary part, it is surely heading off to infinity—the program can move on. But if the program repeats the calculation many times *without* becoming greater than 2, then the point is part of the set. How many times depends on the amount of magnification. For the scales accessible to a personal computer, 100 or 200 is often plenty, and 1,000 is safe.

The program must repeat this process for each of thousands of points on a grid, with a scale that can be adjusted for greater magnification. And the program must display its result. Points in the set can be colored black, other points white. Or for a more vividly appealing picture, the white points can be replaced by colored gradations. If the iteration breaks off after ten repetitions, for example, a program might plot a red dot; for twenty repetitions an orange dot; for forty repetitions a yellow dot, and so on. The choice of colors and cutoff points can be adjusted to suit the programmer's taste. The colors reveal the contours of the terrain just outside the set proper.

of the set's boundary came to the fore. The boundary between two or more attractors in a dynamical system served as a threshold of a kind that seems to govern so many ordinary processes, from the breaking of materials to the making of decisions. Each attractor in such a system has its basin, as a river has a watershed basin that drains into it. Each basin has a boundary. For an influential group in the early 1980s, a most promising new field of mathematics and physics was the study of fractal basin boundaries.

This branch of dynamics concerned itself not with describing the final, stable behavior of a system but with the way a system chooses between competing options. A system like Lorenz's now-classic model has just one attractor in it, one behavior that prevails when the system settles down, and it is a chaotic attractor. Other systems may end up with nonchaotic steady-state behavior—but with more than one possible steady state. The study of fractal basin boundaries was the study of systems that could reach one of several nonchaotic final states, raising the question of how to predict *which*. James Yorke, who pioneered the investigation of fractal basin boundaries a decade after giving chaos its name, proposed an imaginary pinball machine. Like most pinball machines it has a plunger with a spring. You pull back the plunger and release it to send the ball up into the playing area. The machine has the customary tilted landscape of rubber edges and electric bouncers that give the ball a kick of extra energy. The kick is important: it means that energy does not just decay smoothly. For simplicity's sake this machine has no flippers at the bottom, just two exit ramps. The ball must leave by one ramp or the other.

This is deterministic pinball—no shaking the machine. Only one parameter controls the ball's destination, and that is the initial position of the plunger. Imagine that the machine is laid out so that a short pull of the plunger always means that the ball will end up rolling out the right-hand ramp, while a long pull always means that the ball will finish in the left-hand ramp. In between, the behavior gets complex, with the ball bouncing from bumper to bumper in the usual energetic, noisy, and variably long-lived manner before finally choosing one exit or the other.

Now imagine making a graph of the result of each possible starting position of the plunger. The graph is just a line. If a position leads to a right-hand departure, plot a red point, and plot a green point for left. What can we expect to find about these attractors as a function of the initial position?

The boundary proves to be a fractal set, not necessarily self-similar, but infinitely detailed. Some regions of the line will be pure red or green, while others, when magnified, will show new regions of red within the green, or green within the red. For some plunger positions, that is, a tiny change makes no difference. But for others, even an arbitrarily small change will make the difference between red and green.

To add a second dimension meant adding a second parameter, a second degree of freedom. With a pinball machine, for example, one might consider the effect of changing the tilt of the playing slope. One would discover a kind of in-and-out complexity that would give nightmares to engineers responsible for controlling the stability of sensitive, energetic real systems with more than one parameter—electrical power grids, for example, and nuclear generating plants, both of which became targets of chaos-inspired research in the 1980s. For one value of parameter A, parameter B might produce a reassuring, orderly kind of behavior, with coherent regions of stability. Engineers could make studies and graphs of exactly the kind their linear-oriented training suggested. Yet lurking nearby might be another value of parameter A that transforms the importance of parameter B.

Yorke would rise at conferences to display pictures of fractal basin boundaries. Some pictures represented the behavior of forced pendulums that could end up in one of two final states—the forced pendulum being, as his audiences well knew, a fundamental oscillator with many guises in everyday life. "Nobody can say that I've rigged the system by choosing a pendulum," Yorke would say jovially. "This is the kind of thing you see throughout nature. But the behavior is different from anything you see in the literature. It's fractal behavior of a wild kind." The pictures would be fantastic swirls of white and black, as if a kitchen mixing bowl had sputtered a few times in the course of incompletely folding together vanilla and chocolate pudding. To make such pictures, his computer had swept through a 1,000 by 1,000 grid of points, each representing a different starting position for the pendulum, and had plotted the outcome: black or white. These were basins of attraction, mixed and folded by the familiar equations of New-

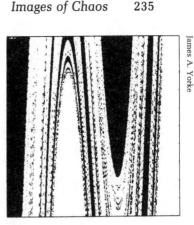

James A. Yorke

FRACTAL BASIN BOUNDARIES. Even when a dynamical system's long-term behavior is not chaotic, chaos can appear at the boundary between one kind of steady behavior and another. Often a dynamical system has more than one equilibrium state, like a pendulum that can come to a halt at either of two magnets placed at its base. Each equilibrium is an attractor, and the boundary between two attractors can be complicated but smooth (*left*). Or the boundary can be complicated but not smooth. The highly fractal interspersing of white and black (*right*) is a phase-space diagram of a pendulum. The system is sure to reach one of two possible steady states. For some starting conditions, the outcome is quite predictable— black is black and white is white. But near the boundary, prediction becomes impossible.

tonian motion, and the result was more boundary than anything else. Typically, more than three-quarters of the plotted points lay on the boundary.

To researchers and engineers, there was a lesson in these pictures—a lesson and a warning. Too often, the potential range of behavior of complex systems had to be guessed from a small set of data. When a system worked normally, staying within a narrow range of parameters, engineers made their observations and hoped that they could extrapolate more or less linearly to less usual behavior. But scientists studying fractal basin boundaries showed that the border between calm and catastrophe could be far more complex than anyone had dreamed. "The whole electrical power grid of the East Coast is an oscillatory system, most of the time stable, and you'd like to know what happens when you perturb it," Yorke said. "You need to know what the boundary is. The fact is, they have no idea what the boundary looks like."

Fractal basin boundaries addressed deep issues in theoretical physics. Phase transitions were matters of thresholds, and Peitgen and Richter looked at one of the best-studied kinds of phase transitions, magnetization and nonmagnetization in materials. Their pictures of such boundaries displayed the peculiarly beautiful complexity that was coming to seem so natural, cauliflower shapes with progressively more tangled knobs and furrows. As they varied the parameters and increased their magnification of details, one picture seemed more and more random, until suddenly, unexpectedly, deep in the heart of a bewildering region, appeared a familiar oblate form, studded with buds: the Mandelbrot set, every tendril and every atom in place. It was another signpost of universality. "Perhaps we should believe in magic," they wrote.

MICHAEL BARNSLEY TOOK a different road. He thought about nature's own images, particularly the patterns generated by living organisms. He experimented with Julia sets and tried other processes, always looking for ways of generating even greater variability. Finally, he turned to randomness as the basis for a new technique of modeling natural shapes. When he wrote about his technique, he called it "the global construction of fractals by means of iterated function systems." When he talked about it, however, he called it "the chaos game."

To play the chaos game quickly, you need a computer with a graphics screen and a random number generator, but in principle a sheet of paper and a coin work just as well. You choose a starting point somewhere on the paper. It does not matter where. You invent two rules, a heads rule and a tails rule. A rule tells you how to take one point to another: "Move two inches to the northeast," or "Move 25 percent closer to the center." Now you start flipping the coin and marking points, using the heads rule when the coin comes up heads and the tails rule when it comes up tails. If you throw away the first fifty points, like a blackjack dealer burying the first few cards in a new deal, you will find the chaos game producing not a random field of dots but a shape, revealed with greater and greater sharpness as the game goes on.

Barnsley's central insight was this: Julia sets and other fractal shapes, though properly viewed as the outcome of a deterministic process, had a second, equally valid existence as the *limit* of a random process. By analogy, he suggested, one could imagine a map of Great Britain drawn in chalk on the floor of a room. A surveyor with standard tools would find it complicated to measure the area of these awkward shapes, with fractal coastlines, after all. But suppose you throw grains of rice into the air one by one, allowing them to fall randomly to the floor and counting the grains that land inside the map. As time goes on, the result begins to approach the area of the shapes—as the limit of a random process. In dynamical terms, Barnsley's shapes proved to be attractors.

The chaos game made use of a fractal quality of certain pictures, the quality of being built up of small copies of the main picture. The act of writing down a set of rules to be iterated randomly captured certain global information about a shape, and the iteration of the rules regurgitated the information without regard to scale. The more fractal a shape was, in this sense, the simpler would be the appropriate rules. Barnsley quickly found that he could generate all the now-classic fractals from Mandelbrot's book. Mandelbrot's technique had been an infinite succession of construction and refinement. For the Koch snowflake or the Sierpiński gasket, one would remove line segments and replace them with specified figures. By using the chaos game instead, Barnsley made pictures that began as fuzzy parodies and grew progressively sharper. No refinement process was necessary: just a single set of rules that somehow embodied the final shape.

Barnsley and his co-workers now embarked on an out-of-control program of producing pictures, cabbages and molds and mud. The key question was how to reverse the process: given a particular shape, how to choose a set of rules. The answer, which he called "collage theorem," was so inanely simple to describe that listeners sometimes thought there must be some trick. You would begin with a drawing of the shape you wanted to reproduce. Barnsley chose a black spleenwort fern for one of his first experiments, having long been a fern buff. Then using a computer terminal and a mouse as pointing device, you would lay small copies over the original shape, letting them overlap sloppily if need be. A highly fractal shape could easily be tiled with copies of itself, a less fractal shape less easily, and at some level of approximation every shape could be tiled.

Michael Barnsley

THE CHAOS GAME. Each new point falls randomly, but gradually the image of a fern emerges. All the necessary information is encoded in a few simple rules

"If the image is complicated, the rules will be complicated," Barnsley said. "On the other hand, if the object has a hidden fractal order to it—and it's a central observation of Benoit's that much of nature does have this hidden order—then it will be possible with a few rules to decode it. The model, then, is more interesteing than a model made with Euclidean geometry, because we know that when you look at the edge of a leaf you don't see straight lines." His first fern, produced with a small desktop computer, perfectly matched the image in the fern book he had since he was a child. "It was a staggering image, correct in every aspect. No biologist would have any trouble identifying it."

In some sense, Barnsley contended, nature must be playing its own version of the chaos game. "There's only so much information in the spore that encodes one fern," he said. "So there's a limit to the elaborateness with which a fern could grow. It's not surprising that we can find equivalent succinct information to describe ferns. It would be surprising if it were otherwise."

But was chance necessary? Hubbard, too, thought about the parallels between the Mandelbrot set and the biological encoding of information, but he bristled at any suggestion that such processes might depend on probability. "There is no randomness in the Mandelbrot set," Hubbard said. "There is no randomness in anything that I do. Neither do I think that the possibility of randomness has any direct relevance to biology. In biology randomness is death, chaos is death. Everything is highly structured. When you clone plants, the order in which the branches come out is exactly the same. The Mandelbrot set obeys an extraordinarily precise scheme leaving nothing to chance whatsoever. I strongly suspect that the day somebody actually figures out how the brain is organized they will discover to their amazement that there is a coding scheme for building the brain which is of extraordinary precision. The idea of randomness in biology is just reflex."

In Barnsley's technique, however, chance serves only as a tool. The results are deterministic and predictable. As points flash across the computer screen, no one can guess where the next one will appear; that depends on the flip of the machine's internal coin. Yet somehow the flow of light always remains within the bounds necessary to carve a shape in phosphorous. To that extent the role of chance is an illusion. "Randomness is a red herring," Barnsley said. "It's central to obtaining images of a certain invariant measure that live upon the fractal object. But the object itself does not depend on the randomness. With probability one, you always draw the same picture.

"It's giving deep information, probing fractal objects with a random algorithm. Just as, when we go into a new room, our eyes dance around it in some order which we might as well take to be random, and we get a good idea of the room. The room is just what it is. The object exists regardless of what I happen to do."

The Mandelbrot set, in the same way, exists. It existed before

Peitgen and Richter began turning it into an art form, before Hubbard and Douady understood its mathematical essence, even before Mandelbrot discovered it. It existed as soon as science created a context—a framework of complex numbers and a notion of iterated functions. Then it waited to be unveiled. Or perhaps it existed even earlier, as soon as nature began organizing itself by means of simple physical laws, repeated with infinite patience and everywhere the same.

The Dynamical
Systems Collective

Communication across the revolutionary divide is inevitably partial.

—THOMAS S. KUHN

SANTA CRUZ was the newest campus in the University of California system, carved into storybook scenery an hour south of San Francisco, and people sometimes said that it looked more like a national forest than a college. The buildings were nestled among redwoods, and, in the spirit of the time, the planners endeavored to leave every tree standing. Little footpaths ran from place to place. The whole campus lay atop a hill, so that every so often you would happen upon the view south across the sparkling waves of Monterey Bay. Santa Cruz opened in 1966, and within a few years it became, briefly, the most selective of the California campuses. Students associated it with many of the icons of the intellectual avant-garde: Norman O. Brown, Gregory Bateson, and Herbert Marcuse lectured there, and Tom Lehrer sang. The school's graduate departments, building from scratch, began with an ambivalent outlook, and physics was no exception. The faculty—about fifteen physicists—was energetic and mostly young, suited to the mix of bright nonconformists attracted to Santa Cruz. They were influenced by the free-thinking ideology of the time; yet they also, the physicists, looked southward toward Caltech and realized that they needed to establish standards and demonstrate their seriousness.

One graduate student whose seriousness no one doubted was Robert Stetson Shaw, a bearded Boston native and Harvard graduate, the oldest of six children of a doctor and a nurse, who in 1977 was about to turn thirty-one years old. That made him a little older than most graduate students, his Harvard career having been interrupted several times for Army service, commune living, and other impromptu experiences somewhere between those extremes. He did not know why he came to Santa Cruz. He had never seen the campus, although he had seen a brochure, with pictures of the redwoods and language about trying new educational philosophies. Shaw was quiet—shy, in a forceful sort of way. He was a good student, and he had reached a point just a few months away from completing his doctoral thesis on superconductivity. No one was particularly concerned that he was wasting time downstairs in the physics building playing with an analog computer.

The education of a physicist depends on a system of mentors and protégés. Established professors get research assistants to help with laboratory work or tedious calculations. In return the graduate students and postdoctoral fellows get shares of their professors' grant money and bits of publication credit. A good mentor helps his student choose problems that will be both manageable and fruitful. If the relationship prospers, the professor's influence helps his protégé find employment. Often their names will be forever linked. When a science does not yet exist, however, few people are ready to teach it. In 1977 chaos offered no mentors. There were no classes in chaos, no centers for nonlinear studies and complex systems research, no chaos textbooks, nor even a chaos journal.

WILLIAM BURKE, A SANTA CRUZ COSMOLOGIST and relativist, ran into his friend Edward A. Spiegel, an astrophysicist, at one o'clock in the morning in the lobby of a Boston hotel, where they were attending a conference on general relativity. "Hey, I've just been listening to the Lorenz attractor," Spiegel said. Spiegel had transmuted this emblem of chaos, using some impromptu circuitry connected to a hi-fi set, into a looping slide-whistle antimelody. He brought Burke into the bar for a drink and explained.

Spiegel knew Lorenz personally, and he had known about chaos since the 1960s. He had made it his business to pursue clues to the possibility of erratic behavior in models of star motion, and he kept in touch with the French mathematicians. Eventually, as a professor at Columbia University, he made turbulence in space—

"cosmic arrhythmias"—the focus of his astronomical study. He had a flair for captivating his colleagues with new ideas, and as the night wore on he captivated Burke. Burke was open to such things. He had made his reputation by working through one of Einstein's more paradoxical gifts to physics, the notion of gravity waves rippling through the fabric of space-time. It was a highly nonlinear problem, with misbehavior related to the troublesome nonlinearities in fluid dynamics. It was also properly abstract and theoretical, but Burke liked down-to-earth physics, too, at one point publishing a paper on the optics of beer glasses: how thick could you make the glass and still leave the appearance of a full portion of beer. He liked to say that he was a bit of a throwback who considered physics to be reality. Furthermore, he had read Robert May's paper in *Nature*, with its plaintive plea for more education about simple nonlinear systems, and he, too, had taken a few hours to play with May's equations on a calculator. So the Lorenz attractor sounded interesting. He had no intention of listening to it. He wanted to see it. When he returned to Santa Cruz, he handed Rob Shaw a piece of paper on which he had scrawled a set of three differential equations. Could Shaw put these on the analog computer?

In the evolution of computers, analog machines represented a blind alley. They did not belong in physics departments, and the existence of such things at Santa Cruz was pure happenstance: the original plans for Santa Cruz had included an engineering school; by the time the engineering school was canceled, an eager purchasing agent had already bought some equipment. Digital computers, built up from circuitry that switched off or on, zero or one, no or yes, gave precise answers to the questions programmers asked, and they proved far more amenable to the miniaturization and acceleration of technology that ruled the computer revolution. Anything done once on a digital computer could be done again, with exactly the same result, and in principle could be done on any other digital computer. Analog computers were, by design, fuzzy. Their building blocks were not yes-no switches but electronic circuits like resistors and capacitors—instantly familiar to anyone who played with radios in the era before solid-state, as Shaw had. The machine at Santa Cruz was a Systron-Donner, a heavy, dusty thing with a patch panel for its front, like the patch panels used by old-fashioned telephone switchboards. Programming the analog computer was a matter of choosing electronic components and plugging cords into the patch panel.

By building up various combinations of circuitry, a programmer simulates systems of differential equations in ways that happen to be well-suited to engineering problems. Say you want to model an automobile suspension with springs, dampers, and mass, to design the smoothest ride. Oscillations in the circuitry can be made to correspond to the oscillations in the physical system. A capacitor takes the place of a spring, inductors represent mass, and so forth. The calculations are not precise. Numerical computation is sidestepped. Instead you have a model made of metal and electrons, quite fast and—best of all—easily adjustable. Simply by turning knobs, you can adjust variables, making the spring stronger or the friction weaker. And you can watch the results change in real time, patterns traced across the screen of an oscilloscope.

Upstairs in the superconductivity laboratory, Shaw was making his desultory way to the end of his thesis work. But he was beginning to spend more and more time playing with the Systron-Donner. He had got far enough to see phase-space portraits of some simple systems—representations of periodic orbits, or limit cycles. If he had seen chaos, in the form of strange attractors, he certainly had not recognized it. The Lorenz equations, handed to him on a piece of paper, were no more complicated than the systems he had been tinkering with. It took just a few hours to patch in the right cords and adjust the knobs. A few minutes later, Shaw knew that he would never finish his superconductivity thesis.

He spent several nights in that basement, watching the green dot of the oscilloscope flying around the screen, tracing over and over the characteristic owl's mask of the Lorenz attractor. The flow of the shape stayed on the retina, a flickering, fluttering thing, unlike any object Shaw's research had shown him. It seemed to have a life of its own. It held the mind just as a flame does, by running in patterns that never repeat. The imprecision and not-quite-repeatability of the analog computer worked to Shaw's advantage. He quickly saw the sensitive dependence on initial conditions that persuaded Edward Lorenz of the futility of long-term weather forecasting. He would set the initial conditions, push the go button, and off the attractor would go. Then he would set the same initial conditions again—as close as physically possible—and the orbit would sail merrily away from its previous course, yet end up on the same attractor.

As a child, Shaw had illusions of what science would be like—dashing off romantically into the unknown. This was finally a kind of exploration that lived up to his illusions. Low-temperature physics was fun from a tinkerer's point of view, with plenty of plumbing and big magnets, liquid helium and dials. But for Shaw it was leading nowhere. Soon he moved the analog computer upstairs, and the room was never used for superconductivity again.

"ALL YOU HAVE TO DO is put your hands on these knobs, and suddenly you are exploring in this other world where you are one of the first travelers and you don't want to come up for air," said Ralph Abraham, a professor of mathematics who dropped by in the early days to watch the Lorenz attractor in motion. He had been with Steve Smale in the most glorious early days at Berkeley, and so he was one of very few members of the Santa Cruz faculty with a background that would let him grasp the significance of Shaw's game-playing. His first reaction was astonishment at the speed of the display—and Shaw pointed out that he was using extra capacitors to keep it from running even faster. The attractor was robust, too. The imprecision of the analog circuitry proved that—the tuning and tweaking of knobs did not make the attractor vanish, did not turn it into something random, but turned it or bent it in ways that slowly began to make sense. "Rob had the spontaneous experience where a little exploration reveals all the secrets," Abraham said. "All the important concepts—the Lyapunov exponent, the fractal dimension—would just naturally occur to you. You would see it and start exploring."

Was this science? It certainly was not mathematics, this computer work with no formalisms or proofs, and no amount of sympathetic encouragement from people like Abraham could change that. The physics faculty saw no reason to think it was physics, either. Whatever it was, it drew an audience. Shaw usually left his door open, and it happened that the entrance to the physics department was just across the hall. The foot traffic was considerable. Before long, he found himself with company.

The group that came to call itself the Dynamical Systems Collective—others sometimes called it the Chaos Cabal—depended on Shaw as its quiet center. He suffered from a certain diffidence in putting his ideas forward in the academic marketplace; fortunately for him, his new associates had no such problem. They, meanwhile, often returned to his steady vision of how to carry out an unplanned program of exploring an unrecognized science.

Doyne Farmer, a tall, angular, and sandy-haired Texas native, became the group's most articulate spokesman. In 1977 he was twenty-four years old, all energy and enthusiasm, a machine for ideas. Those who met him sometimes suspected at first that he was all hot air. Norman Packard, three years younger, a boyhood friend who had grown up in the same New Mexico town, Silver City, arrived at Santa Cruz that fall, just as Farmer was beginning a year off to devote all his energy to his plan for applying the laws of motion to the game of roulette. This enterprise was as earnest as it was far-fetched. For more than a decade Farmer and a changing cast of fellow physicists, professional gamblers, and hangers-on pursued the roulette dream. Farmer did not give it up even after he joined the Theoretical Division of Los Alamos National Laboratory. They calculated tilts and trajectories, wrote and rewrote custom software, embedded computers in shoes and made nervous forays into gambling casinos. But nothing quite worked as planned. At one time or another, all the members of the collective but Shaw lent their energy to roulette, and it had to be said that the project gave them unusual training in the rapid analysis of dynamical systems, but it did little to reassure the Santa Cruz physics faculty that Farmer was taking science seriously.

The fourth member of the group was James Crutchfield, the youngest and the only native Californian. He was short and powerfully built, a stylish windsurfer and, most important for the collective, an instinctive master of computing. Crutchfield came to Santa Cruz as an undergraduate, worked as a laboratory assistant on Shaw's pre-chaos superconductivity experiments, spent a year commuting "over the hill," as they said in Santa Cruz, to a job at

IBM's research center in San Jose, and did not actually join the physics department as a graduate student until 1980. By then he had spent two years hanging around Shaw's laboratory and rushing to pick up the mathematics he needed to understand dynamical systems. Like the rest of the group, he left the department's standard track behind.

It was spring in 1978 before the department quite believed that Shaw was abandoning his superconductivity thesis. He was so close to finishing. No matter how bored he was, the faculty reasoned that he could rush through the formalities, get his doctorate and move on to the real world. As for chaos, there were questions of academic suitability. No one at Santa Cruz was qualified to supervise a course of study in this field-without-a-name. No one had ever received a doctorate in it. Certainly no jobs were available for graduates with this kind of specialty. There was also the matter of money. Physics at Santa Cruz, as at every American university, was financed mostly by the National Science Foundation and other agencies of the federal government through research grants to members of the faculty. The Navy, the Air Force, the Department of Energy, the Central Intelligence Agency—all dispensed vast sums for pure research, without necessarily caring about immediate application to hydrodynamics, aerodynamics, energy, or intelligence. A faculty physicist would get enough to pay for laboratory equipment and the salaries of research assistants—graduate students, who would piggy-back themselves on his grant. He would pay for their photocopying, for their travel to meetings, even for salaries to keep them going in the summers. Otherwise a student was financially adrift. This was the system from which Shaw, Farmer, Packard, and Crutchfield now cut themselves off.

When certain kinds of electronic equipment began to disappear at night, it became prudent to look for them in Shaw's former low-temperature laboratory. Occasionally a member of the collective would be able to cadge a hundred dollars from the graduate student association, or the physics department would find a way to appropriate that much. Plotters, converters, electronic filters began to accumulate. A particle physics group down the hall had a small digital computer that was destined for the scrapheap; it found its way to Shaw's lab. Farmer became a particular specialist in scrounging computer time. One summer he was invited to the National Center for Atmospheric Research in Boulder, Colorado, where huge computers handle research on such tasks as global weather modeling, and his ability to siphon expensive time from these machines stunned the climatologists.

The Santa Cruzians' tinkering sensibility served them well. Shaw had grown up "gizmo-oriented." Packard had fixed television sets as a boy in Silver City. Crutchfield belonged to the first generation of mathematicians for whom the logic of computer processors was a natural language. The physics building itself, in its shady redwood setting, was like physics buildings everywhere, with a universal ambience of cement floors and walls that always needed repainting, but the room taken over by the chaos group developed its own atmosphere, with piles of papers and pictures of Tahitian islanders on the walls and, eventually, printouts of strange attractors. At almost any hour, though night was a safer bet than morning, a visitor could see members of the group rearranging circuitry, yanking out patch cords, arguing about consciousness or evolution, adjusting an oscilloscope display, or just staring while a glowing green spot traced a curve of light, its orbit flickering and seething like something alive.

"THE SAME THING REALLY DREW all of us: the notion that you could have determinism but not really," Farmer said. "The idea that all these classical deterministic systems we'd learned about could generate randomness was intriguing. We were driven to understand what made that tick.

"You can't appreciate the kind of revelation that is unless you've been brainwashed by six or seven years of a typical physics curriculum. You're taught that there are classical models where everything is determined by initial conditions, and then there are quantum mechanical models where things are determined but you have to contend with a limit on how much initial information you can gather. *Nonlinear* was a word that you only encountered in the back of the book. A physics student would take a math course and the last chapter would be on nonlinear equations. You would usually skip that, and, if you didn't, all they would do is take these nonlinear equations and reduce them to linear equations, so you just get approximate solutions anyway. It was just an exercise in frustration.

"We had no concept of the real difference that nonlinearity makes in a model. The idea that an equation could bounce around in an apparently random way—that was pretty exciting. You would say, 'Where is this random motion coming from? I don't see it in the equations.' It seemed like something for nothing, or something out of nothing."

Crutchfield said, "It was a realization that here is a whole realm of physical experience that just doesn't fit in the current framework. Why wasn't that part of what we were taught? We had a chance to look around the immediate world—a world so mundane it was wonderful—and understand something."

They enchanted themselves and dismayed their professors with leaps to questions of determinism, the nature of intelligence, the direction of biological evolution.

"The glue that held us together was a long-range vision," Packard said, "It was striking to us that if you take regular physical systems which have been analyzed to death in classical physics, but you take one little step away in parameter space, you end up with something to which all of this huge body of analysis does not apply.

"The phenomenon of chaos could have been discovered long, long ago. It wasn't, in part because this huge body of work on the dynamics of regular motion didn't lead in that direction. But if you just look, there it is. It brought home the point that one should allow oneself to be guided by the physics, by observations, to see what kind of theoretical picture one could develop. In the long run we saw the investigation of complicated dynamics as an entry point that might lead to an understanding of really, really complicated dynamics."

Farmer said, "On a philosophical level, it struck me as an operational way to define free will, in a way that allowed you to reconcile free will with determinism. The system is deterministic, but you can't say what it's going to do next. At the same time, I'd always felt that the important problems out there in the world had to do with the creation of organization, in life or intelligence. But how did you study that? What biologists were doing seemed so applied and specific; chemists certainly weren't doing it; mathe-

maticians weren't doing it at all, and it was something that physicists just didn't do. I always felt that the spontaneous emergence
of self-organization ought to be part of physics.

"Here was one coin with two sides. Here was order, with
randomness emerging, and then one step further away was randomness with its own underlying order."

SHAW AND HIS COLLEAGUES had to turn their raw enthusiasm
into a scientific program. They had to ask questions that could be
answered and that would be worth answering. They sought ways
of connecting theory and experiment—there, they felt, was a gap
that needed to be closed. Before they could even begin, they had
to learn what was known and what was not, and this itself was a
formidable challenge.

They were hindered by the tendency of communication to
travel piecemeal in science, particularly when a new subject jumps
across the established subdisciplines. Often they had no idea
whether they were on new or old territory. One invaluable antidote
to their ignorance was Joseph Ford, an advocate of chaos at the
Georgia Institute of Technology. Ford had already decided that
nonlinear dynamics was the future of physics—the entire future—
and had set himself up as a clearinghouse of information on journal articles. His background was in nondissipative chaos, the chaos
of astronomical systems or of particle physics. He had an unusually intimate knowledge of the work being done by the Soviet
school, and he made it his business to seek out connections with
anyone who remotely shared the philosophical spirit of this new
enterprise. He had friends everywhere. Any scientist who sent in
a paper on nonlinear science would have his work summarized
on Ford's growing list of abstracts. The Santa Cruz students found
out about Ford's list and made up a form postcard for requesting
prepublication copies of articles. Soon the preprints flooded in.

They realized that many sorts of questions could be asked
about strange attractors. What are their characteristic shapes? What
is their topological structure? What does the geometry reveal about
the physics of the related dynamical systems? The first approach
was the hands-on exploration that Shaw began with. Much of
the mathematical literature dealt directly with structure, but the

mathematical approach struck Shaw as too detailed—still too many trees and not enough forest. As he worked his way through the literature, he felt that the mathematicians, deprived by their own traditions of the new tools of computing, had been buried in the particular complexities of orbit structures, infinities here and discontinuities there. The mathematicians had not cared especially about analog fuzziness—from the physicist's point of view, the fuzziness that surely controlled real-world systems. Shaw saw on his oscilloscope not the individual orbits but an envelope in which the orbits were embedded. It was the envelope that changed as he gently turned the knobs. He could not give a rigorous explanation of the folds and twists in the language of mathematical topology. Yet he began to feel that he understood them.

A physicist wants to make measurements. What was there in these elusive moving images to measure? Shaw and the others tried to isolate the special qualities that made strange attractors so enchanting. *Sensitive dependence on initial conditions*—the tendency of nearby trajectories to pull away from one another. This was the quality that made Lorenz realize that deterministic long-term weather forecasting was an impossibility. But where were the calipers to gauge such a quality? Could unpredictability itself be measured?

The answer to this question lay in a Russian conception, the Lyapunov exponent. This number provided a measure of just the topological qualities that corresponded to such concepts as unpredictability. The Lyapunov exponents in a system provided a way of measuring the conflicting effects of stretching, contracting, and folding in the phase space of an attractor. They gave a picture of all the properties of a system that lead to stability or instability. An exponent greater than zero meant stretching—nearby points would separate. An exponent smaller than zero meant contraction. For a fixed-point attractor, all the Lyapunov exponents were negative, since the direction of pull was inward toward the final steady state. An attractor in the form of a periodic orbit had one exponent of exactly zero and other exponents that were negative. A strange attractor, it turned out, had to have at least one positive Lyapunov exponent.

To their chagrin, the Santa Cruz students did not invent this idea, but they developed it in the most practical ways possible,

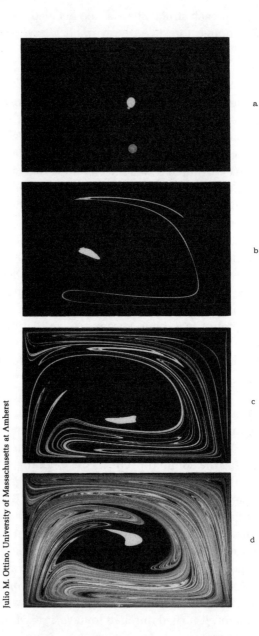

CHAOTIC MIXING. One blob mixes rapidly; another blob, just a bit closer to the center, barely mixes at all. In experiments by Julio M. Ottino and others with real fluids, the process of mixing—ubiquitous in nature and industry, yet still poorly understood—proved intimately bound up with the mathematics of chaos. The patterns revealed a stretching and folding that led back to the horseshoe map of Smale.

learning how to measure Lyapunov exponents and relate them to other important properties. They used computer animation to make movies illustrating the beating together of order and chaos in dynamical systems. Their analysis showed vividly how some systems could create disorder in one direction while remaining trim and methodical in another. One movie showed what happened to a tiny cluster of nearby points—representing initial conditions—on a strange attractor as the system evolved in time. The cluster began to spread out and lose focus. It turned into a dot and then a blob. For certain kinds of attractors, the blob would quickly spread all over. Such attractors were efficient at *mixing*. For other attractors, though, the spreading would only occur in certain directions. The blob would become a band, chaotic along one axis and orderly along another. It was as if the system had an orderly impulse and a disorderly one together, and they were decoupling. As one impulse led to random unpredictability, the other kept time like a precise clock. Both impulses could be defined and measured.

THE MOST CHARACTERISTICALLY Santa Cruzian imprint on chaos research involved a piece of mathematics cum philosophy known as information theory, invented in the late 1940s by a researcher at the Bell Telephone Laboratories, Claude Shannon. Shannon called his work "The Mathematical Theory of Communication," but it concerned a rather special quantity called information, and the name information theory stuck. The theory was a product of the electronic age. Communication lines and radio transmissions were carrying a certain thing, and computers would soon be storing this same thing on punch cards or magnetic cylinders, and the thing was neither knowledge nor meaning. Its basic units were not ideas or concepts or even, necessarily, words or numbers. This thing could be sense or nonsense—but the engineers and mathematicians could measure it, transmit it, and test the transmission for accuracy. *Information* proved as good a word as any, but people had to remember that they were using a specialized value-free term without the usual connotations of facts, learning, wisdom, understanding, enlightenment.

Hardware determined the shape of the theory. Because information was stored in binary on-off switches newly designated as bits, bits became the basic measure of information. From a technical point of view, information theory became a handle for grasping how noise in the form of random errors interfered with the flow of bits. It gave a way of predicting the necessary carrying capacity of communication lines or compact disks or any technology that encoded language, sounds, or images. It offered a theoretical means of reckoning the effectiveness of different schemes for correcting errors—for example, using some bits as checks on others. It put teeth into the crucial notion of "redundancy." In terms of Shannon's information theory, ordinary language contains greater than fifty percent redundancy in the form of sounds or letters that are not strictly necessary to conveying a message. This is a familiar idea; ordinary communication in a world of mumblers and typographical errors depends on redundancy. The famous advertisement for shorthand training—*if u cn rd ths msg . . .* —illustrated the point, and information theory allowed it to be measured. Redundancy is a predictable departure from the random. Part of the redundancy in ordinary language lies in its meaning, and that part is hard to quantify, depending as it does on people's shared knowledge of their language and the world. This is the part that allows people to solve crossword puzzles or fill in the missing word at the end of a. But other kinds of redundancy lend themselves more easily to numerical measures. Statistically, the likelihood that any letter in English will be "e" is far greater than one in twenty-six. Furthermore, letters do not have to be counted as isolated units. Knowing that one letter in an English text is "t" helps in predicting that the next might be "h" or "o," and knowing two letters helps even more, and so on. The statistical tendency of various two- and three-letter combinations to turn up in a language goes a long way toward capturing some characteristic essence of the language. A computer guided only by the relative likelihood of the possible sequences of three letters can produce an otherwise random stream of nonsense that is recognizably *English* nonsense. Cryptologists have long made use of such statistical patterns in breaking simple codes. Communications engineers now use them in devising techniques to compress data, removing the redundancy to save space on a transmission line or storage disk. To Shannon, the right way to look at such patterns was this: a stream of data in ordinary language is less than random; each new bit is partly constrained by the bits before; thus each new bit carries somewhat less than a bit's worth of real information. There was a hint of paradox floating in this formulation. The more random a data stream, the more information would be conveyed by each new bit.

Beyond its technical aptness to the beginning of the computer era, Shannon information theory gained a modest philosophical stature, and a surprising part of the theory's appeal to people beyond Shannon's field could be attributed to the choice of a single word: entropy. As Warren Weaver put it in a classic exposition of information theory, "When one meets the concept of entropy in communication theory, he has a right to be rather excited—a right to suspect that one has hold of something that may turn out to be basic and important." The concept of entropy comes from thermodynamics, where it serves as an adjunct of the Second Law, the inexorable tendency of the universe, and any isolated system in it, to slide toward a state of increasing disorder. Divide a swimming pool in half with some barrier; fill one half with water and one with ink; wait for all to be still; lift the barrier; simply through the random motion of molecules, eventually the ink and water will mix. The mixing never reverses itself, even if you wait till the end of the universe, which is why the Second Law is so often said to be the part of physics that makes time a one-way street. Entropy is the name for the quality of systems that increases under the Second Law—mixing, disorder, randomness. The concept is easier to grasp intuitively than to measure in any real-life situation. What would be a reliable test for the level of mixing of two substances? One could imagine counting the molecules of each in some sample. But what if they were arranged yes–no–yes–no–yes–no–yes–no? Entropy could hardly be described as high. One could count just the even molecules, but what if the arrangement were yes–no–no–yes–yes–no–no–yes? Order intrudes in ways that defy any straightforward counting algorithm. And in information theory, issues of meaning and representation present extra complications. A sequence like 01 0100 0100 0010 111 010 11 00 000 0010 111 010 11 0100 0 000 000 . . . might seem orderly only to an observer familiar with Morse code and Shakespeare. And what about the topologically perverse patterns of a strange attractor?

To Robert Shaw, strange attractors were engines of information. In his first and grandest conception, chaos offered a natural way of returning to the physical sciences, in reinvigorated form, the ideas that information theory had drawn from thermodynamics. Strange attractors, conflating order and disorder, gave a challenging twist to the question of measuring a system's entropy. Strange attractors served as efficient mixers. They created unpredictability. They raised entropy. And as Shaw saw it, they created information where none existed.

Norman Packard was reading *Scientific American* one day and spotted an advertisement for an essay contest called the Louis Jacot competition. This was suitably far-fetched—a prize lucratively endowed by a French financier who had nurtured a private theory about the structure of the universe, galaxies within galaxies. It called for essays on Jacot's theme, whatever that was. ("It sounded like a bunch of crank mail," Farmer said.) But judging the competition was an impressive panel drawn from France's scientific establishment, and the money was impressive as well. Packard showed the advertisement to Shaw. The deadline was New Year's Day 1978.

By now the collective was meeting regularly in an outsized old Santa Cruz house not far from the beach. The house accumulated flea-market furniture and computer equipment, much of which was devoted to the roulette problem. Shaw kept a piano there, on which he would play baroque music or improvise his own blend of the classical and modern. In their meetings the physicists developed a working style, a routine of throwing out ideas and filtering them through some sieve of practicality, reading the literature, and conceiving papers of their own. Eventually they learned to collaborate on journal articles in a reasonably efficient round-robin way, but the first paper was Shaw's, one of the few he would produce, and he kept the writing of it to himself, characteristically. Also characteristically, it was late.

In December 1977 Shaw headed out from Santa Cruz to attend the first meeting of the New York Academy of Sciences devoted to chaos. His superconductivity professor paid his fare, and Shaw arrived uninvited to hear in person the scientists he knew only from their writing. David Ruelle. Robert May. James Yorke. Shaw was awed by these men and also by the astronomical $35 room charge at the Barbizon Hotel. Listening to the talks, he swung back and forth between feeling that he had been ignorantly reinventing ideas that these men had worked out in considerable detail and, on the other hand, feeling that he had an important new point of view to contribute. He had brought the unfinished draft of his information theory paper, scribbled in longhand on scraps of paper in a folder, and he tried unsuccessfully to get a typewriter, first from the hotel and then from local repair shops. In the end he took his folder away with him. Later, when his friends begged him for details, he told them the high point had been a dinner in honor of Edward Lorenz, who was finally receiving the recognition that had eluded him for so many years. When Lorenz walked into the room, shyly holding his wife's hand, the scientists rose to their feet to give him an ovation. Shaw was struck by how terrified the meteorologist looked.

A few weeks later, on a trip to Maine, where his parents had a vacation house, he finally mailed his paper to the Jacot competition. New Year's had passed, but the envelope was generously backdated by the local postmaster. The paper—a blend of esoteric mathematics and speculative philosophy, illustrated with cartoon-like drawings by Shaw's brother Chris—won an honorable mention. Shaw received a large enough cash prize to pay for a trip to Paris to collect the honor. It was a small enough achievement, but it came at a difficult moment in the group's relations with the department. They desperately needed whatever external signs of credibility they could find. Farmer was giving up astrophysics, Packard was abandoning statistical mechanics, and Crutchfield still was not ready to call himself a graduate student. The department felt matters were out of control.

"STRANGE ATTRACTORS, CHAOTIC BEHAVIOR, and Information Flow" circulated that year in a preprint edition that eventually reached about 1,000, the first painstaking effort to weave together information theory and chaos.

Shaw brought some assumptions of classical mechanics out

of the shadows. Energy in natural systems exists on two levels: the macroscales, where everyday objects can be counted and measured, and the microscales, where countless atoms swim in random motion, unmeasurable except as an average entity, temperature. As Shaw noted, the total energy living in the microscales could outweigh the energy of the macroscales, but in classical systems this thermal motion was irrelevant—isolated and unusable. The scales do not communicate with one another. "One does not have to know the temperature to do a classical mechanics problem," he said. It was Shaw's view, however, that chaotic and near-chaotic systems bridged the gap between macroscales and microscales. Chaos was the creation of information.

One could imagine water flowing past an obstruction. As every hydrodynamicist and white-water canoeist knows, if the water flows fast enough, it produces whorls downstream. At some speed, the whorls stay in place. At some higher speed, they move. An experimenter could choose a variety of methods for extracting data from such a system, with velocity probes and so forth, but why not try something simple: pick a point directly downstream from the obstruction and, at uniform time intervals, ask whether the whorl is to the right or the left.

If the whorls are static, the data stream will look like this: left–left–left–left–left–left–left–left–left–left–left–left–left–left–left–left–left–left–left–left–. After a while, the observer starts to feel that new bits of data are failing to offer new information about the system.

Or the whorls might be moving back and forth periodically: left–right–left–right–left–right–left–right–left–right–left–right–left–right–left–right–left–right–left–right–. Again, though at first the system seems one degree more interesting, it quickly ceases to offer any surprises.

As the system becomes chaotic, however, strictly by virtue of its unpredictability, it generates a steady stream of information. Each new observation is a new bit. This is a problem for the experimenter trying to characterize the system completely. "He could never leave the room," as Shaw said. "The flow would be a continuous source of information."

Where is this information coming from? The heat bath of the microscales, billions of molecules in their random thermodynamic dance. Just as turbulence transmits energy from large scales downward through chains of vortices to the dissipating small scales of viscosity, so information is transmitted back from the small scales to the large—at any rate, this was how Shaw and his colleagues began describing it. And the channel transmitting the information upward is the strange attractor, magnifying the initial randomness just as the Butterfly Effect magnifies small uncertainties into large-scale weather patterns.

The question was how much. Shaw found—after unwittingly duplicating some of their work—that again Soviet scientists had been there first. A. N. Kolmogorov and Yasha Sinai had worked out some illuminating mathematics for the way a system's "entropy per unit time" applies to the geometric pictures of surfaces stretching and folding in phase space. The conceptual core of the technique was a matter of drawing some arbitrarily small box around some set of initial conditions, as one might draw a small square on the side of a balloon, then calculating the effect of various expansions or twists on the box. It might stretch in one direction, for example, while remaining narrow in the other. The change in area corresponded to an introduction of uncertainty about the system's past, a gain or loss of information.

To the extent that information was just a fancy word for unpredictability, this conception merely matched the ideas that such scientists as Ruelle were developing. But the information theory framework allowed the Santa Cruz group to adopt a body of mathematical reasoning that had been well investigated by communications theorists. The problem of adding extrinsic noise to an otherwise deterministic system, for example, was new in dynamics but familiar enough in communications. The real appeal for these young scientists, however, was only partly the mathematics. When they spoke of systems generating information, they thought about the spontaneous generation of pattern in the world. "At the pinnacle of complicated dynamics are processes of biological evolution, or thought processes," Packard said. "Intuitively there seems a clear sense in which these ultimately complicated systems are generating information. Billions of years ago there were just blobs of protoplasm; now billions of years later here we are. So information has been created and stored in our structure. In the development of one person's mind from childhood, information is clearly not just accumulated but also generated—created from connections that were not there before." It was the kind of talk that could make a sober physicist's head spin.

THEY WERE TINKERERS FIRST, though, and philosophers only second. Could they make a bridge from the strange attractors they knew so well to the experiments of classical physics? It was one thing to say that right–left–right–right–left–right–left–left–left– right was unpredictable and information-generating. It was quite another to take a stream of real data and measure its Lyapunov exponent, its entropy, its dimension. Still, the Santa Cruz physicists had made themselves more comfortable with these ideas than had any of their older colleagues. By living with strange attractors day and night, they convinced themselves that they recognized them in the flapping, shaking, beating, swaying phenomena of their everyday lives.

They had a game they would play, sitting at a coffeehouse. They would ask: How far away is the nearest strange attractor? Was it that rattling automobile fender? That flag snapping erratically in a steady breeze? A fluttering leaf? "You don't see something until you have the right metaphor to let you perceive it," Shaw said, echoing Thomas S. Kuhn. Before long, their relativist friend Bill Burke was quite convinced that the speedometer in his car was rattling in the nonlinear fashion of a strange attractor. And Shaw, settling on an experimental project that would occupy him for years to come, adopted as homely a dynamical system as any physicist could imagine: a dripping faucet. Most people imagine the canonical dripping faucet as relentlessly periodic, but it is not necessarily so, as a moment of experimentation reveals. "It's a simple example of a system that goes from predictable behavior to unpredictable behavior," Shaw said. "If you turn it up a little bit, you can see a regime where the pitter-patter is irregular. As it turns out, it's not a predictable pattern beyond a short time. So even something as simple as a faucet can generate a pattern that is eternally creative."

As a generator of organization, the dripping faucet offers little to work with. It generates only drips, and each drip is about the same as the last. But for a beginning investigator of chaos, the dripping faucet proved to have certain advantages. Everyone already has a mental picture of it. The data stream is as one-dimensional as could be: a rhythmic drumbeat of single points measured in time. None of these qualities could be found in systems that the Santa Cruz group explored later—the human immune system, for example, or the troublesome beam-beam effect that was inexplicably degrading the performance of colliding particle beams at the Stanford Linear Accelerator Center to the north. Experimenters like Libchaber and Swinney obtained a one-dimensional data stream by placing a probe arbitrarily at one point in a slightly more complex system. In the dripping faucet the single line of data is all there is. And it isn't even a continuously varying velocity or temperature—just a list of drip times.

Asked to organize an attack on such a system, a traditional physicist might begin by making as complete a physical model as possible. The processes governing the creation and breaking off of drips are understandable, if not quite so simple as they might seem. One important variable is the rate of flow. (This had to be slow compared to most hydrodynamic systems. Shaw usually looked at drop rates of 1 to 10 per second, which meant a flow rate of 30 to 300 gpf—gallons per fortnight.) Other variables include the viscosity of the fluid and the surface tension. A drop of water hanging from a faucet, waiting to break off, assumes a complicated three-dimensional shape, and the calculation of this shape alone was, as Shaw said, "a state-of-the-art computer calculation." Furthermore, the shape is far from static. A drop filling with water is like a little elastic bag of surface tension, oscillating this way and that, gaining mass and stretching its walls until it passes a critical point and snaps off. A physicist trying to model the drip problem completely—writing down sets of coupled nonlinear partial differential equations with appropriate boundary conditions and then trying to solve them—would find himself lost in a deep, deep thicket.

An alternative approach would be to forget about the physics and look only at the data, as though it were coming out of a black box. Given a list of numbers representing intervals between drips, could an expert in chaotic dynamics find something useful to say? Indeed, as it turned out, methods could be devised for organizing such data and working backward into the physics, and these methods became critical to the applicability of chaos to real-world problems.

But Shaw began halfway between these extremes, by making a sort of caricature of a complete physical model. Ignoring drop shapes, ignoring complex motions in three dimensions, he roughly summarized drip physics. He imagined a weight hanging from a spring. He imagined that the weight grew steadily with time. As it grew, the spring would stretch and the weight would hang lower and lower. When it reached a certain point, a portion of the weight would break off. The amount that would detach, Shaw supposed arbitrarily, would depend strictly on the speed of the descending weight when it reached the cutoff point.

Then, of course, the remaining weight would bounce back up, as springs do, with oscillations that graduate students learn to model using standard equations. The interesting feature of the model—the *only* interesting feature, and the nonlinear twist that made chaotic behavior possible—was that the next drip depended on how the springiness interacted with the steadily increasing weight. A down bounce might help the weight reach the cutoff point that much sooner, or an up bounce might delay the process slightly. With a real faucet, drops are not all the same size. The size depends both on the velocity of the flow and on the direction of the bounce. If a drop starts off its life already moving downward, then it will break off sooner. If it happens to be on the rebound, it will be able to fill with a bit more water before it snaps. Shaw's model was exactly crude enough to be summed up in three differential equations, the minimum necessary for chaos, as Poincaré and Lorenz had shown. But would it generate as much complexity as a real faucet? And would the complexity be of the same kind?

Thus Shaw found himself sitting in a laboratory in the physics building, a big plastic tub of water over his head, a tube running down to a premium-quality hardware-store brass nozzle. As each drop fell, it interrupted a light beam, and a microcomputer in the next room recorded the time. Meanwhile Shaw had his three arbitrary equations up and running on the analog computer, producing a stream of imaginary data. One day he performed some show-and-tell for the faculty—a "pseudocolloquium," as Crutchfield said, because graduate students were not permitted to give formal colloquiums. Shaw played a tape of a faucet making its drumbeat on a piece of tin. And he had his computer going click-click-click in a crisp syncopation, revealing patterns to the ear. He had solved the problem simultaneously from front and back, and his listeners could hear the deep structure in this seemingly disorderly system. But to go further, the group needed a way of taking raw data from any experiment and working backward to equations and strange attractors that characterized chaos.

With a more complicated system, one could imagine plotting one variable against another, relating changes in temperature or velocity to the passage of time. But the dripping faucet provided only a series of times. So Shaw tried a technique that may have been the Santa Cruz group's cleverest and most enduring practical contribution to the progress of chaos. It was a method of reconstructing a phase space for an unseen strange attractor, and it could be applied to any series of data at all. For the dripping faucet data, Shaw made a two-dimensional graph in which the x axis represented a time interval between a pair of drops and the y axis represented the next time interval. If 150 milliseconds passed between drop one and drop two, and then 150 milliseconds passed between drop two and drop three, he would plot a point at the position 150–150.

That was all there was to it. If the dripping was regular, as it tended to be when the water flowed slowly and the system was in its "water clock regime," the graph would be suitably dull. Every point would land at the same place. The graph would be a single dot. Or almost. Actually, the first difference between the computer dripping faucet and the real dripping faucet was that the real version was subject to noise, and exceedingly sensitive. "It turns out that the thing is an excellent seismometer," Shaw said ironically, "very efficient in bringing noise up from the little-league scales to the big-league scales." Shaw ended up doing most of his work at night, when foot traffic in the physics corridors was lightest. Noise meant that, instead of the single dot predicted by theory, he would see a slightly fuzzy blob.

As the flow rate was increased, the system would go through a period-doubling bifurcation. Drops would fall in pairs. One interval might be 150 milliseconds, and the next might be 80. So the graph would show two fuzzy blobs, one centered at 150–80 and the other at 80–150. The real test came when the pattern became chaotic. If it were truly random, points would be scattered all over the graph. There would be no relation to be found between one interval and the next. But if a strange attractor were hidden in the data, it might reveal itself as a coalescence of fuzziness into distinguishable structures.

Often three dimensions were necessary to see the structure, but that was no problem. The technique could easily be generalized to higher-dimensional graph-making. Instead of plotting interval n against interval $n + 1$, one could plot interval n against interval $n + 1$ against interval $n + 2$. It was a trick—a gimmick. Ordinarily a three-dimensional graph required knowledge of three independent variables in a system. The trick gave three variables for the price of one. It reflected the faith of these scientists that order was so deeply ingrained in apparent disorder that it would find a way of expressing itself even to experimenters who did not know which physical variables to measure or who were not able to measure such variables directly. As Farmer said, "When you think about a variable, the evolution of it must be influenced by whatever other variables it's interacting with. Their values must somehow be contained in the history of that thing. Somehow their mark must be there." In the case of Shaw's dripping faucet the pictures illustrated the point. In three dimensions, especially, the patterns emerged, resembling loopy trails of smoke left by an out-of-control sky-writing plane. Shaw was able to match plots of the experimental data with data produced by his analog computer model, the main difference being that the real data was always fuzzier, smeared out by noise. Even so, the structure was unmistakable. The Santa Cruz group began collaborating with such experienced experimentalists as Harry Swinney, who had moved to the University of Texas in Austin, and they learned how to retrieve strange attractors from all kinds of systems. It was a matter of embedding the data in a phase space of enough dimensions. Soon Floris Takens, who had invented strange attractors with David Ruelle, independently gave a mathematical foundation for this powerful technique of reconstructing the phase space of an attractor from a stream of real data. As countless researchers soon discovered, the technique distinguishes between mere noise and chaos, in the new sense: orderly disorder created by simple processes. Truly random data remains spread out in an undefined mess. But chaos—deterministic and patterned—pulls the data into visible shapes. Of all the possible pathways of disorder, nature favors just a few.

THE TRANSITION FROM REBEL to physicist was slow. Every so often, sitting in a coffeehouse or working in their laboratory, one or another of the students would have to fight back amazement that their scientific fantasy had not ended. *God, we're still doing this and it still makes sense*, as Jim Crutchfield would say. *We're still here. How far is it going to go?*

Their chief supporters on the faculty were the Smale protégé Ralph Abraham in the mathematics department and in the physics department Bill Burke, who had himself made "czar of the analog computer" to protect the collective's claim to this piece of equipment, at least. The rest of the physics faculty found itself in a more difficult position. A few years later, some professors denied bitterly that the collective had been forced to overcome indifference or opposition from the department. The collective reacted just as bitterly to what it considered revisionist history on the part of belated converts to chaos. "We had no advisor, nobody telling us what to do," said Shaw. "We were in an adversary role for years, and it continues to this day. We were never funded at Santa Cruz. Every one of us worked for considerable periods of time without pay, and it was a shoestring operation the entire way, with no intellectual or other guidance."

By its lights, though, the faculty tolerated and even abetted a long period of research that seemed to fall short of any substantial kind of science. Shaw's thesis advisor in superconductivity kept him on salary for a year or so, long after Shaw had veered away from low-temperature physics. No one ever quite ordered the chaos research to stop. At worst the faculty reached an attitude of benevolent discouragement. Each member of the collective was taken aside from time to time for heart-to-heart talks. They were warned that, even if somehow a way could be found to justify doctorates, no one would be able to help the students find jobs in a nonexistent field. This may be a fad, the faculty would say, and then where will you be? Yet outside the redwood shelter of the Santa Cruz hills, chaos was creating its own scientific establishment, and the Dynamical Systems Collective had to join it.

One year Mitchell Feigenbaum came by, making the rounds of the lecture circuit to explain his breakthrough in universality. As always, his talks were abstrusely mathematical; renormalization group theory was an esoteric piece of condensed matter physics that these students had not studied. Besides, the collective was more interested in real systems than in delicate one-dimensional maps. Doyne Farmer, meanwhile, heard that a Berkeley mathematician, Oscar E. Lanford III, was exploring chaos, and he went up to talk. Lanford listened politely and then looked at Farmer and said they had nothing in common. He was trying to understand Feigenbaum.

How deadly! Where's the guy's sense of scope? Farmer thought. "He was looking at these little orbits. Meanwhile we were into information theory with all its profundity, taking chaos apart, seeing what make it tick, trying to relate metric entropy and Lyapunov exponents to more statistical measures."

In his conversation with Farmer, Lanford did not emphasize universality, and only later did Farmer realize that he had missed the point. "It was my naïveté," Farmer said. "The idea of universality was not just a great result. Mitchell's thing was also a technique to employ a whole army of unemployed critical phenomena people.

"Up to that point it appeared that nonlinear systems would have to be treated in a case-by-case way. We were trying to come up with a language to quantify it and describe it, but it still seemed as though everything would have to be treated case by case. We saw no way to put systems in classes and write solutions that would be valid for the whole class, as in linear systems. Universality meant finding properties that were exactly the same in quantifiable ways for everything in that class. *Predictable* properties. That's why it was really important.

"And there was a sociological factor that pumped even more fuel. Mitchell cast his results in the language of renormalization. He took all this machinery that people in critical phenomena had been skilled in using. Those guys were having a hard time, because there didn't seem to be any interesting problems left for them to do. They were looking around for something else to apply their bag of tricks to. And suddenly Feigenbaum came forward with his extremely significant application of this bag of tricks. It spawned an entire subdiscipline."

Quite independently, however, the Santa Cruz students began to make an impression of their own. Within the department their star began to rise after a surprise appearance at a midwinter meeting in condensed matter physics in Laguna Beach in 1978, organized by Bernardo Huberman of the Xerox Palo Alto Research Center and Stanford University. The collective was not invited, but it went nonetheless, bundling itself into Shaw's 1959 Ford ranch-style station wagon, an automobile known as the Cream Dream. Just in case, the group brought some equipment, including a huge television monitor and a videotape. When an invited speaker canceled at the last minute, Huberman invited Shaw to take his place. The timing was perfect. Chaos had attained the status of buzzword, but few of the physicists attending the conference knew what it meant. So Shaw began by explaining attractors in phase space: first fixed points (where everything stops); then limit cycles (where everything oscillates); then strange attractors (everything else). He demonstrated with his computer graphics on videotape. ("Audiovisual aids gave us an edge," he said. "We could hypnotize them with flashing lights.") He illuminated the Lorenz attractor and the dripping faucet. He explained the geometry—how shapes are stretched and folded, and what that meant in the grand terms of information theory. And for good measure, he put in a few words at the end about shifting paradigms. The talk was a popular triumph, and in the audience were several members of the Santa Cruz faculty, seeing chaos for the first time through the eyes of their colleagues.

IN 1979 THE WHOLE GROUP attended the second chaos meeting of the New York Academy of Sciences, this time as participants, and now the field was exploding. The 1977 meeting had been Lorenz's, attended by specialists numbering in the dozens. This meeting was Feigenbaum's, and scientists came by the hundreds. Where two years earlier Rob Shaw had shyly tried to find a typewriter so that he could produce a paper to leave under people's doors, now the Dynamical Systems Collective had become a virtual printing press, producing papers rapidly and always under joint authorship.

But the collective could not go on forever. The closer it came to the real world of science, the closer it came to unraveling. One day Bernardo Huberman called. He asked for Rob Shaw, but he happened to get Crutchfield. Huberman needed a collaborator for a tight, simple paper about chaos. Crutchfield, the youngest member of the collective, concerned about being thought of as merely its "hacker," was beginning to realize that in one respect the Santa Cruz faculty had been right all along: each of the students was someday going to have to be judged as an individual. Huberman, furthermore, had all the sophistication about the profession of physics that the students lacked, and in particular he knew how to get the most mileage from a given piece of work. He had his doubts, having seen their laboratory—"It was all very vague, you know, sofas and bean bags, like stepping into a time machine, flower children and the 1960s again." But he needed an analog computer, and in fact Crutchfield managed to get his research program running in hours. The collective was a problem, though. "All the guys want in," Crutchfield said at one point, and Huberman said absolutely not. "It's not just the credit, it's the blame. Suppose the paper is wrong—you're going to blame a collective? I'm not part of a *collective.*" He wanted one partner for a clean job.

The result was just what Huberman had hoped for: the first paper about chaos to be published in the premier American journal for reporting breakthroughs in physics, *Physical Review Letters.* In terms of scientific politics this was a nontrivial achievement. "To us it was fairly obvious stuff," Crutchfield said, "but what Bernardo understood was that it would have a huge impact." It was also one beginning of the group's assimilation into the real world. Farmer was angered, seeing in Crutchfield's defection an undermining of the collective spirit.

Crutchfield was not alone in stepping outside the group. Soon Farmer himself, and Packard, too, were collaborating with established physicists and mathematicians: Huberman, Swinney, Yorke. The ideas formed in the cauldron at Santa Cruz became a firm part of the framework of the modern study of dynamical systems. When a physicist with a mass of data wanted to investigate its dimension or its entropy, the appropriate definitions and working techniques might well be those created in the years of patching plugs in the Systron-Donner analog computer and staring at the oscilloscope. Climate specialists would argue about whether the chaos of the world's atmosphere and oceans had infinite dimensions, as traditional dynamicists would assume, or somehow followed a low-dimensional strange attractor. Economists analyzing stock market data would try to find attractors of dimension 3.7 or 5.3. The lower the dimension, the simpler the system. Many mathematical peculiarities had to be sorted and understood. Fractal dimension, Hausdorff dimension, Lyapunov dimension, information dimension—the subtleties of these measures of a chaotic system were best explained by Farmer and Yorke. An attractor's dimension was "the first level of knowledge necessary to characterize its properties." It was the feature that gave "the amount of information necessary to specify the position of a point on the attractor to within a given accuracy." The methods of the Santa Cruz students and their older collaborators tied these ideas to the other important measures of systems: the rate of decay of predictability, the rate of information flow, the tendency to create mixing. Sometimes scientists using these methods would find themselves plotting data, drawing little boxes, and counting the number of data points in each box. Yet even such seemingly crude techniques brought chaotic systems for the first time within the reach of scientific understanding.

Meanwhile, having learned to look for strange attractors in flapping flags and rattling speedometers, the scientists made a point of finding the symptoms of deterministic chaos all through the current literature of physics. Unexplained noise, surprising fluctuations, regularity mixing with irregularity—these effects popped up in papers from experimentalists working with everything from particle accelerators to lasers to Josephson junctions. The chaos specialists would make these symptoms their own, telling the unconverted, in effect, your problems are our problems. "Several experiments on Josephson junction oscillators have revealed a striking noise-rise phenomena," a paper would begin, "which cannot be accounted for in terms of thermal fluctuations."

By the time the collective departed, some of the Santa Cruz faculty had turned to chaos, too. Other physicists, though, felt in retrospect that Santa Cruz had missed an opportunity to begin the kind of national center for work in nonlinear dynamics that soon began appearing on other campuses. In the early 1980s the members of the collective graduated and dispersed. Shaw finished his dissertation in 1980, Farmer in 1981, Packard in 1982. Crutchfield's appeared in 1983, a typographical hodgepodge interleaving typed pages with no less than eleven papers already published in the journals of physics and mathematics. He went on to the University of California at Berkeley. Farmer joined the Theoretical Division of Los Alamos. Packard and Shaw joined the Institute for Advanced Study in Princeton. Crutchfield studied video feedback loops. Farmer worked on "fat fractals" and modeled the complex dynamics of the human immune system. Packard explored spatial chaos and the formation of snowflakes. Only Shaw seemed reluctant to join the mainstream. His own influential legacy comprised just two papers, one that had won him a trip to Paris and one, about the dripping faucet, that summed up all his Santa Cruz research. Several times, he came close to quitting science altogether. As one of his friends said, he was oscillating.

Inner Rhythms

The sciences do not try to explain, they hardly even try to interpret, they mainly make models. By a model is meant a mathematical construct which, with the addition of certain verbal interpretations, describes observed phenomena. The justification of such a mathematical construct is solely and precisely that it is expected to work.

—JOHN VON NEUMANN

BERNARDO HUBERMAN LOOKED OUT over his audience of assorted theoretical and experimental biologists, pure mathematicians and physicians and psychiatrists, and he realized that he had a communication problem. He had just finished an unusual talk at an unusual gathering in 1986, the first major conference on chaos in biology and medicine, under the various auspices of the New York Academy of Sciences, the National Institute of Mental Health, and the Office of Naval Research. In the cavernous Masur Auditorium at the National Institutes of Health outside Washington, Huberman saw many familiar faces, chaos specialists of long standing, and many unfamiliar ones as well. An experienced speaker could expect some audience impatience—it was the conference's last day, and it was dangerously close to lunch time.

Huberman, a dapper black-haired Californian transplanted from Argentina, had kept up his interest in chaos since his collaborations with members of the Santa Cruz gang. He was a research fellow at the Xerox Corporation's Palo Alto Research Center. But sometimes he dabbled in projects that did not belong to the corporate mission, and here at the biology conference he had just finished describing one of those: a model for the erratic eye movement of schizophrenics.

Psychiatrists have struggled for generations to define schizophrenia and classify schizophrenics, but the disease has been almost as difficult to describe as to cure. Most of its symptoms appear in mind and behavior. Since 1908, however, scientists have known of a physical manifestation of the disease that seems to afflict not only schizophrenics but also their relatives. When patients try to watch a slowly swinging pendulum, their eyes cannot track the smooth motion. Ordinarily the eye is a remarkably smart instrument. A healthy person's eyes stay locked on moving targets without the least conscious thought; moving images stay frozen in place on the retina. But a schizophrenic's eyes jump about disruptively in small increments, overshooting or undershooting the target and creating a constant haze of extraneous movements. No one knows why.

Physiologists accumulated vast amounts of data over the years, making tables and graphs to show the patterns of erratic eye motion. They generally assumed that the fluctuations came from fluctuations in the signal from the central nervous system controlling the eye's muscles. Noisy output implied noisy input, and perhaps some random disturbances afflicting the brains of schizophrenics were showing up in the eyes. Huberman, a physicist, assumed otherwise and made a modest model.

He thought in the crudest possible way about the mechanics of the eye and wrote down an equation. There was a term for the amplitude of the swinging pendulum and a term for its frequency. There was a term for the eye's inertia. There was a term for damping, or friction. And there were terms for error correction, to give the eye a way of locking in on the target.

As Huberman explained to his audience, the resulting equation happens to describe an analogous mechanical system: a ball rolling in a curved trough while the trough swings from side to side. The side-to-side motion corresponds to the motion of the pendulum, and the walls of the trough correspond to the error-correcting feature, tending to push the ball back toward the center. In the now-standard style of exploring such equations, Huberman had run his model for hours on a computer, changing the various parameters and making graphs of the resulting behaviors. He found both order and chaos. In some regimes, the eye would track smoothly; then, as the degree of nonlinearity was increased, the system would go through a fast period-doubling sequence and produce a kind of disorder that was indistinguishable from the disorder reported in the medical literature.

In the model, the erratic behavior had nothing to do with any outside signal. It was an inevitable consequence of too much nonlinearity in the system. To some of the doctors listening, Huberman's model seemed to match a plausible genetic model for schizophrenia. A nonlinearity that could either stabilize the system or disrupt it, depending on whether the nonlinearity was weak or strong, might correspond to a single genetic trait. One psychiatrist compared the concept to the genetics of gout, in which too high a level of uric acid creates pathological symptoms. Others, more familiar than Huberman with the clinical literature, pointed out that schizophrenics were not alone; a whole range of eye movement problems could be found in different kinds of neurological patients. Periodic oscillations, aperiodic oscillations, all sorts of dynamical behavior could be found in the data by anyone who cared to go back and apply the tools of chaos.

But for every scientist present who saw new lines of research opening up, there was another who suspected Huberman of grossly oversimplifying his model. When it came time for questions, their annoyance and frustration spilled out. "My problem is, what guides you in the modeling?" one of these scientists said. "Why look for these specific elements of nonlinear dynamics, namely these bifurcations and chaotic solutions?"

Huberman paused. "Oh, okay. Then I truly failed at stating the purpose of this. The model is simple. Someone comes to me and says, we see this, so what do you think happens. So I say, well, what is the possible explanation. So they say, well, the only thing we can come up with is something that is fluctuating over such a short time in your *head*. So then I say, well look, I'm a chaotician of sorts, and I know that the simplest nonlinear tracking model you can write down, the *simplest*, has these generic features, regardless of the details of what these things are like. So I do that and people say, oh, that's very interesting, we never thought that perhaps this was intrinsic chaos in the system.

"The model does not have any neurophysiological data that I can even defend. All I'm saying is that the *simplest* tracking is something that tends to make an error and go to zero. That's the way we move our eyes, and that's the way an antenna tracks an airplane. You can apply this model to anything."

Out on the floor, another biologist took the microphone, still frustrated by the stick-figure simplicity of Huberman's model. In real eyes, he pointed out, four muscle-control systems operate simultaneously. He began a highly technical description of what he considered realistic modeling, explaining how, for example, the mass term is thrown away because the eye is heavily over-damped. "And there's one additional complication, which is that the amount of mass present depends on the velocity of rotation, because part of the mass lags behind when the eye accelerates very rapidly. The jelly inside the eye lags behind when the outer casing rotates very fast."

Pause. Huberman was stymied. Finally one of the conference organizers, Arnold Mandell, a psychiatrist with a long interest in chaos, took the microphone from him.

"Look, as a shrink I want to make an interpretation. What you've just seen is what happens when a nonlinear dynamicist working with low-dimensional global systems comes to talk to a biologist who's been using mathematical tools. The idea that in fact there are universal properties of systems, built into the *simplest* representations, alienates all of us. So the question is 'What is the subtype of the schizophrenia,' 'There are four ocular motor systems,' and 'What is the modeling from the standpoint of the actual physical structure,' and it begins to decompose.

"What's actually the case is that, as physicians or scientists learning all 50,000 parts of everything, we resent the possibility that there are in fact universal elements of motion. And Bernardo comes up with one and look what happens."

Huberman said, "It happened in physics five years ago, but by now they're convinced."

THE CHOICE IS ALWAYS the same. You can make your model more complex and more faithful to reality, or you can make it simpler and easier to handle. Only the most naïve scientist believes that the perfect model is the one that perfectly represents reality. Such a model would have the same drawbacks as a map as large and detailed as the city it represents, a map depicting every park, every street, every building, every tree, every pothole, every inhabitant, and every map. Were such a map possible, its specificity would defeat its purpose: to generalize and abstract.

Mapmakers highlight such features as their clients choose. Whatever their purpose, maps and models must simplify as much as they mimic the world.

For Ralph Abraham, the Santa Cruz mathematician, a good model is the "daisy world" of James E. Lovelock and Lynn Margulis, proponents of the so-called Gaia hypothesis, in which the conditions necessary for life are created and maintained by life itself in a self-sustaining process of dynamical feedback. The daisy world is perhaps the simplest imaginable version of Gaia, so simple as to seem idiotic. "Three things happen," as Abraham put it, "white daisies, black daisies, and unplanted desert. Three colors: white, black, and red. How can this teach us anything about our planet? It explains how temperature regulation emerges. It explains why this planet is a good temperature for life. The daisy world model is a terrible model, but it teaches how biological homeostasis was created on earth."

White daisies reflect light, making the planet cooler. Black daisies absorb light, lowering the albedo, or reflectivity, and thus making the planet warmer. But white daisies "want" warm weather, meaning that they thrive preferentially as temperatures rise. Black daisies want cool weather. These qualities can be expressed in a set of differential equations and the daisy world can be set in motion on a computer. A wide range of initial conditions will lead to an equilibrium attractor—and not necessarily a static equilibrium.

"It's just a mathematical model of a conceptual model, and that's what you want—you don't *want* high-fidelity models of biological or social systems," Abraham said. "You just put in the albedos, make some initial planting, and watch billions of years of evolution go by. And you educate children to be better members of the board of directors of the planet."

The paragon of a complex dynamical system and to many scientists, therefore, the touchstone of any approach to complexity is the human body. No object of study available to physicists offers such a cacophony of counterrhythmic motion on scales from macroscopic to microscopic: motion of muscles, of fluids, of currents, of fibers, of cells. No physical system has lent itself to such an obsessive brand of reductionism: every organ has its own microstructure and its own chemistry, and student physiologists spend years just on the naming of parts. Yet how ungraspable these parts can be! At its most tangible, a body part can be a seemingly well-defined organ like the liver. Or it can be a spatially challenging network of solid and liquid like the vascular system. Or it can be an invisible assembly, truly as abstract a thing as "traffic" or "democracy," like the immune system, with its lymphocytes and T4 messengers, a miniaturized cryptography machine for encoding and decoding data about invading organisms. To study such systems without a detailed knowledge of their anatomy and chemistry would be futile, so heart specialists learn about ion transport through ventricular muscle tissue, brain specialists learn the electrical particulars of neuron firing, and eye specialists learn the name and place and purpose of each ocular muscle. In the 1980s chaos brought to life a new kind of physiology, built on the idea that mathematical tools could help scientists understand global complex systems independent of local detail. Researchers increasingly recognized the body as a place of motion and oscillation—and they developed methods of listening to its variegated drumbeat. They found rhythms that were invisible on frozen microscope slides or daily blood samples. They studied chaos in respiratory disorders. They explored feedback mechanisms in the control of red and white blood cells. Cancer specialists speculated about periodicity and irregularity in the cycle of cell growth. Psychiatrists explored a multidimensional approach to the prescription of antidepressant drugs. But surprising findings about one organ dominated the rise of this new physiology, and that was the heart, whose animated rhythms, stable or unstable, healthy or pathological, so precisely measured the difference between life and death.

EVEN DAVID RUELLE HAD STRAYED from formalism to speculate about chaos in the heart—"a dynamical system of vital interest to every one of us," he wrote.

"The normal cardiac regime is periodic, but there are many nonperiodic pathologies (like ventricular fibrillation) which lead to the steady state of death. It seems that great medical benefit might be derived from computer studies of a realistic mathematical model which would reproduce the various cardiac dynamical regimes."

Teams of researchers in the United States and Canada took up the challenge. Irregularities in the heartbeat had long since been discovered, investigated, isolated, and categorized. To the trained ear, dozens of irregular rhythms can be distinguished. To the trained eye, the spiky patterns of the electrocardiogram offer clues to the source and the seriousness of an irregular rhythm. A layman can gauge the richness of the problem from the cornucopia of names available for different sorts of arrhythmias. There are ectopic beats, electrical alternans, and torsades de pointes. There are high-grade block and escape rhythms. There is parasystole (atrial or ventricular, pure or modulated). There are Wenckebach rhythms (simple or complex). There is tachycardia. Most damaging of all to the prospect for survival is fibrillation. This naming of rhythms, like the naming of parts, comforts physicians. It allows specificity in diagnosing troubled hearts, and it allows some intelligence to bear on the problem. But researchers using the tools of chaos began to discover that traditional cardiology was making the wrong generalizations about irregular heartbeats, inadvertently using superficial classifications to obscure deep causes.

They discovered the dynamical heart. Almost always their backgrounds were out of the ordinary. Leon Glass of McGill University in Montreal was trained in physics and chemistry, where he indulged an interest in numbers and in irregularity, too, completing his doctoral thesis on atomic motion in liquids before turning to the problem of irregular heartbeats. Typically, he said, specialists diagnose many different arrhythmias by looking at short strips of electrocardiograms. "It's treated by physicians as a pattern recognition problem, a matter of identifying patterns they have seen before in practice and in textbooks. They really don't analyze in detail the dynamics of these rhythms. The dynamics are much richer than anybody would guess from reading the textbooks."

At Harvard Medical School, Ary L. Goldberger, co-director of the arrhythmia laboratory of Beth Israel Hospital in Boston, believed that the heart research represented a threshold for collaboration between physiologists and mathematicians and physicists. "We're at a new frontier, and a new class of phenomenology is out there," he said. "When we see bifurcations, abrupt changes in behavior, there is nothing in conventional linear models to account for that. Clearly we need a new class of models, and physics seems to provide that." Goldberger and other scientists had to overcome barriers of scientific language and institutional classification. A considerable obstacle, he felt, was the uncomfortable antipathy of many physiologists to mathematics. "In 1986 you won't find the word fractals in a physiology book," he said. "I think in 1996 you won't be able to find a physiology book without it."

A doctor listening to the heartbeat hears the whooshing and pounding of fluid against fluid, fluid against solid, and solid against solid. Blood courses from chamber to chamber, squeezed by the contracting muscles behind, and then stretches the walls ahead. Fibrous valves snap shut audibly against the backflow. The muscle contractions themselves depend on a complex three-dimensional wave of electrical activity. Modeling any one piece of the heart's behavior would strain a supercomputer; modeling the whole interwoven cycle would be impossible. Computer modeling of the kind that seems natural to a fluid dynamics expert designing airplane wings for Boeing or engine flows for the National Aeronautics and Space Administration is an alien practice to medical technologists.

Trial and error, for example, has governed the design of artificial heart valves, the metal and plastic devices that now prolong the lives of those whose natural valves wear out. In the annals of engineering a special place must be reserved for nature's own heart valve, a filmy, pliant, translucent arrangement of three tiny parachute-like cups. To let blood into the heart's pumping chambers, the valve must fold gracefully out of the way. To keep blood from backing up when the heart pumps it forward, the valve must fill and slam closed under the pressure, and it must do so, without leaking or tearing, two or three billion times. Human engineers have not done so well. Artificial valves, by and large, have been borrowed from plumbers: standard designs like "ball in cage," tested, at great cost, in animals. To overcome the obvious problems of leakage and stress failure was hard enough. Few could have anticipated how hard it would be to eliminate another problem. By changing the patterns of fluid flow in the heart, artificial valves create areas of turbulence and areas of stagnation; when blood stagnates, it forms clots; when clots break off and travel to the brain, they cause strokes. Such clotting was the fatal barrier to making artificial hearts. Only in the mid-1980s, when mathematicians at the Courant Institute of New York University applied new computer modeling techniques to the problem, did the design of heart valves begin to take full advantage of available technology. Their computer made motion pictures of a beating heart, two-dimensional but vividly recognizable. Hundreds of dots, representing particles of blood, stream through the valve, stretching the elastic walls of the heart and creating whirling vortices. The mathematicians found that the heart adds a whole level of complexity to the standard fluid flow problem, because any realistic model must take into account the elasticity of the heart walls themselves. Instead of flowing over a rigid surface, like air over an airplane wing, blood changes the heart surface dynamically and nonlinearly.

Even subtler, and far deadlier, was the problem of arrhythmias. Ventricular fibrillation causes hundreds of thousands of sudden deaths each year in the United States alone. In many of those cases, fibrillation has a specific, well-known trigger: blockage of the arteries, leading to the death of the pumping muscle. Cocaine use, nervous stress, hypothermia—these, too, can predispose a person to fibrillation. In many cases the onset of fibrillation remains mysterious. Faced with a patient who has survived an attack of fibrillation, a doctor would prefer to see damage—evidence of a cause. A patient with a seemingly healthy heart is actually more likely to suffer a new attack.

There is a classic metaphor for the fibrillating heart: a bag of worms. Instead of contracting and relaxing, contracting and relaxing in a repetitive, periodic way, the heart's muscle tissue writhes, uncoordinated, helpless to pump blood. In a normally beating heart the electrical signal travels as a coordinated wave through the three-dimensional structure of the heart. When the signal arrives, each cell contracts. Then each cell relaxes for a critical refractory period, during which it cannot be set off again prematurely. In a fibrillating heart the wave breaks up. The heart is never all contracted or all relaxed.

One perplexing feature of fibrillation is that many of the heart's individual components can be working normally. Often the heart's pacemaking nodes continue to send out regular electrical ticks. Individual muscle cells respond properly. Each cell receives its stimulus, contracts, passes the stimulus on, and relaxes to wait for the next stimulus. In autopsy the muscle tissue may reveal no damage at all. That is one reason chaos experts believed that a new, global approach was necessary: the parts of a fibrillating heart seem to be working, yet the whole goes fatally awry. Fibrillation is a disorder of a complex system, just as mental disorders—whether or not they have chemical roots—are disorders of a complex system.

The heart will not stop fibrillating on its own. This brand of chaos is stable. Only a jolt of electricity from a defibrillation device—a jolt that any dynamicist recognizes as a massive perturbation—can return the heart to its steady state. On the whole, defibrillators are effective. But their design, like the design of artificial heart valves, has required much guesswork. "The business of determining the size and shape of that jolt has been strictly empirical," said Arthur T. Winfree, a theoretical biologist. "There just hasn't been any theory about that. It now appears that some assumptions are not correct. It appears that defibrillators can be radically redesigned to improve their efficiency many fold and therefore improve the chance of success many fold." For other abnormal heart rhythms an assortment of drug therapies have been tried, also based largely on trial and error—"a black art," as Winfree put it. Without a sound theoretical understanding of the heart's dynamics, it is tricky to predict the effects of a given drug. "A wonderful job has been done in the last twenty years of finding out all the nitty gritty details of membrane physiology, all the detailed, precise workings of the immense complexity of all the parts of the heart. That essential part of the business is in good shape. What's gotten overlooked is the other side, trying to achieve some global perspective on how it all works."

WINFREE CAME FROM A FAMILY in which no one had gone to college. He got started, he would say, by not having a proper education. His father, rising from the bottom of the life insurance business to the level of vice president, moved the family almost yearly up and down the East Coast, and Winfree attended more than a dozen schools before finishing high school. He developed a feeling that the interesting things in the world had to do with biology and mathematics and a companion feeling that no standard combination of the two subjects did justice to what was interesting. So he decided not to take a standard approach. He took a five-year course in engineering physics at Cornell University, learning applied mathematics and a full range of hands-on laboratory styles. Prepared to be hired into the military-industrial complex, he got a doctorate in biology, striving to combine experiment with theory in new ways. He began at Johns Hopkins, left because of conflicts with the faculty, continued at Princeton, left because of conflicts with the faculty there, and finally was awarded a Princeton degree from a distance, when he was already teaching at the University of Chicago.

Winfree is a rare kind of thinker in the biological world, bringing a strong sense of geometry to his work on physiological problems. He began his exploration of biological dynamics in the early seventies by studying biological clocks—circadian rhythms. This was an area traditionally governed by a naturalist's approach: this rhythm goes with that animal, and so forth. In Winfree's view the problem of circadian rhythms should lend itself to a mathematical style of thinking. "I had a headful of nonlinear dynamics and realized that the problem could be thought of, and ought to be thought of, in those qualitative terms. Nobody had any idea what the mechanisms of biological clocks are. So you have two choices. You can wait until the biochemists figure out the mechanism of clocks and then try to derive some behavior from the known mechanisms, or you can start studying how clocks work in terms of complex systems theory and nonlinear and topological dynamics. Which I undertook to do."

At one time he had a laboratory full of mosquitoes in cages. As any camper could guess, mosquitoes perk up around dusk each day. In a laboratory, with temperature and light kept constant to shield them from day and night, mosquitoes turn out to have an inner cycle of not twenty-four hours but twenty-three. Every twenty-three hours, they buzz around with particular intensity. What keeps them on track outdoors is the jolt of light they get each day; in effect, it resets their clock.

Winfree shined artificial light on his mosquitoes, in doses that he carefully regulated. These stimuli either advanced or delayed the next cycle, and he plotted the effect against the timing of the blast. Then, instead of trying to guess at the biochemistry involved, he looked at the problem topologically—that is, he looked at the qualitative shape of the data, instead of the quantitative details. He came to a startling conclusion: There was a *singularity* in the geometry, a point different from all the other points. Looking at the singularity, he predicted that one special, precisely timed burst of light would cause a complete breakdown of a mosquito's biological clock, or any other biological clock.

The prediction was surprising, but Winfree's experiments bore it out. "You go to a mosquito at midnight and give him a certain number of photons, and that particularly well-timed jolt turns off the mosquito's clock. He's an insomniac after that—he'll doze, buzz for a while, all at random, and he'll continue doing that for as long as you care to watch, or until you come along with another jolt. You've given him perpetual jet lag." In the early seventies Winfree's mathematical approach to circadian rhythms stirred little general interest, and it was hard to extend the laboratory technique to species that would object to sitting in little cages for months at a time.

Human jet lag and insomnia remain on the list of unsolved problems in biology. Both bring out the worst charlatanism—useless pills and nostrums. Researchers did amass data on human subjects, usually students or retired people, or playwrights with plays to finish, willing to accept a few hundred dollars a week to live in "time isolation": no daylight, no temperature change, no clocks, and no telephones. People have a sleep-wake cycle and also a body-temperature cycle, both nonlinear oscillators that restore themselves after slight perturbations. In isolation, without a daily resetting stimulus, the temperature cycle seems to be about twenty-five hours, with the low occurring during sleep. But experiments by German researchers found that after some weeks the sleep-wake cycle would detach itself from the temperature cycle and become erratic. People would stay awake for twenty or thirty hours at a time, followed by ten or twenty hours of sleep. Not only would the subjects remain unaware that their day had lengthened, they would refuse to believe it when told. Only in the mid-1980s, though, did researchers begin to apply Winfree's systematic ap-

CHEMICAL CHAOS. Waves propagating outward in concentric circles and even spiral waves were signs of chaos in a widely studied chemical reaction, the Beluzov–Zhabotinsky reaction. Similar patterns have been observed in dishes of millions of amoeba. Arthur Winfree theorized that such waves are analogous to the waves of electrical activity coursing through heart muscles, regularly or erratically.

proach to humans, starting with an elderly woman who did needle-point in the evening in front of banks of bright light. Her cycle changed sharply, and she reported feeling great, as if she were driving in a car with the top down. As for Winfree, he had moved on to the subject of rhythms in the heart.

Actually, he would not have said "moved on." To Winfree it was the same subject—different chemistry, same dynamics. He had gained a specific interest in the heart, however, after he help-lessly witnessed the sudden cardiac deaths of two people, one a relative on a summer vacation, the other a man in a pool where Winfree was swimming. Why should a rhythm that has stayed on track for a lifetime, two billion or more uninterrupted cycles, through relaxation and stress, acceleration and deceleration, sud-denly break into an uncontrolled, fatally ineffectual frenzy?

WINFREE TOLD THE STORY of an early researcher, George Mines, who in 1914 was twenty-eight years old. In his laboratory at McGill University in Montreal, Mines made a small device capable of delivering small, precisely regulated electrical impulses to the heart.

"When Mines decided it was time to begin work with human beings, he chose the most readily available experimental subject: himself," Winfree wrote. "At about six o'clock that evening, a janitor, thinking it was unusually quiet in the laboratory, entered the room. Mines was lying under the laboratory bench surrounded by twisted electrical equipment. A broken mechanism was at-tached to his chest over the heart and a piece of apparatus nearby was still recording the faltering heartbeat. He died without re-covering consciousness."

One might guess that a small but precisely timed shock can throw the heart into fibrillation, and indeed even Mines had guessed it, shortly before his death. Other shocks can advance or retard the next beat, just as in circadian rhythms. But one difference between hearts and biological clocks, a difference that cannot be set aside even in a simplified model, is that a heart has a shape in space. You can hold it in your hand. You can track an electrical wave through three dimensions.

To do so, however, requires ingenuity. Raymond E. Ideker of

Duke University Medical Center read an article by Winfree in *Scientific American* in 1983 and noted four specific predictions about inducing and halting fibrillation based on nonlinear dynamics and topology. Ideker didn't really believe them. They seemed too speculative and, from a cardiologist's point of view, so abstract. Within three years, all four had been tested and confirmed, and Ideker was conducting an advanced program to gather the richer data necessary to develop the dynamical approach to the heart. It was, as Winfree said, "the cardiac equivalent of a cyclotron."

The traditional electrocardiogram offers only a gross one-dimensional record. During heart surgery a doctor can take an electrode and move it from place to place on the heart, sampling as many as fifty or sixty sites over a ten-minute period and thus producing a sort of composite picture. During fibrillation this technique is useless. The heart changes and quivers far too rapidly. Ideker's technique, depending heavily on real-time computer processing, was to embed 128 electrodes in a web that he would place over a heart like a sock on a foot. The electrodes recorded the voltage field as each wave traveled through the muscle, and the computer produced a cardiac map.

Ideker's immediate intention, beyond testing Winfree's theoretical ideas, was to improve the electrical devices used to halt fibrillation. Emergency medical teams carry standard versions of defibrillators, ready to deliver a strong DC shock across the thorax of a stricken patient. Experimentally, cardiologists have developed a small implantable device to be sewn inside the chest cavity of patients thought to be especially at risk, although identifying such patients remains a challenge. An implantable defibrillator, somewhat bigger than a pacemaker, sits and waits, listening to the steady heartbeat, until it becomes necessary to release a burst of electricity. Ideker began to assemble the physical understanding necessary to make the design of defibrillators less a high-priced guessing game, more a science.

WHY SHOULD THE LAWS of chaos apply to the heart, with its peculiar tissue—cells forming interconnected branching fibers, transporting ions of calcium, potassium, and sodium? That was the question puzzling scientists at McGill and the Massachusetts Institute of Technology.

Leon Glass and his colleagues Michael Guevara and Alvin Schrier at McGill carried out one of the most talked-about lines of research in the whole short history of nonlinear dynamics. They used tiny aggregates of heart cells from chicken embryos seven days old. These balls of cells, 1/200 of an inch across, placed in a dish and shaken together, began beating spontaneously at rates on the order of once a second, with no outside pacemaker at all. The pulsation was clearly visible through a microscope. The next step was to apply an external rhythm as well, and the McGill scientists did this through a microelectrode, a thin tube of glass drawn out to a fine point and inserted into one of the cells. An electric potential was passed through the tube, stimulating the cells with a strength and a rhythm that could be adjusted at will.

They summed up their findings this way in *Science* in 1981: "Exotic dynamic behavior that was previously seen in mathematical studies and in experiments in the physical sciences may in general be present when biological oscillators are periodically perturbed." They saw period-doubling—beat patterns that would bifurcate and bifurcate again as the stimulus changed. They made Poincaré maps and circle maps. They studied intermittency and mode-locking. "Many different rhythms can be established between a stimulus and a little piece of chicken heart," Glass said. "Using nonlinear mathematics, we can understand quite well the different rhythms and their orderings. Right now, the training of cardiologists has almost no mathematics, but the way we are looking at these problems is the way that at some point in the future people will have to look at these problems."

Meanwhile, in a joint Harvard-M.I.T. program in health sciences and technology, Richard J. Cohen, a cardiologist and a physicist, found a range of period-doubling sequences in experiments with dogs. Using computer models, he tested one plausible scenario, in which the wavefront of electrical activity breaks up on islands of tissue. "It is a clear instance of the Feigenbaum phenomenon," he said, "a regular phenomenon which, under certain circumstances, becomes chaotic, and it turns out that the electrical activity in the heart has many parallels with other systems that develop chaotic behavior."

The McGill scientists also went back to old data accumulated on different kinds of abnormal heartbeats. In one well-known syndrome, abnormal, ectopic beats are interspersed with normal, sinus beats. Glass and his colleagues examined the patterns, counting the numbers of sinus beats between ectopic beats. In some people, the numbers would vary, but for some reason they would always be odd: 3 or 5 or 7. In other people, the number of normal beats would always be part of the sequence: 2, 5, 8, 11. . . .

"People have made these weird numerology observations, but the mechanisms are not very easy to understand," Glass said. "There is often some type of regularity in these numbers, but there is often great irregularity also. It's one of the slogans in this business: order in chaos."

Traditionally, thoughts about fibrillation took two forms. One classic idea was that secondary pacemaking signals come from abnormal centers within the heart muscle itself, conflicting with the main signal. These tiny ectopic centers fire out waves at uncomfortable intervals, and the interplay and overlapping has been thought to break up the coordinated wave of contraction. The research by the McGill scientists provided some support for this idea, by demonstrating that a full range of dynamical misbehavior can arise from the interplay between an external pulse and a rhythm inherent in the heart tissue. But why secondary pacemaking centers should develop in the first place has remained hard to explain.

The other approach focused not on the initiation of electrical waves but on the way they are conducted geographically through the heart, and the Harvard-M.I.T. researchers remained closer to this tradition. They found that abnormalities in the wave, spinning in tight circles, could cause "re-entry," in which some areas begin a new beat too soon, preventing the heart from pausing for the quiet interval necessary to maintain coordinated pumping.

By stressing the methods of nonlinear dynamics, both groups of researchers were able to use the awareness that a small change in one parameter—perhaps a change in timing or electrical conductivity—could push an otherwise healthy system across a bifurcation point into a qualitatively new behavior. They also began to find common ground for studying heart problems globally, linking disorders that were previously considered unrelated. Furthermore, Winfree believed that, despite their different focus, both the ectopic beat school and the re-entry school were right. His topological approach suggested that the two ideas might be one and the same.

"Dynamical things are generally counterintuitive, and the heart is no exception," Winfree said. Cardiologists hoped that the research would lead to a scientific way of identifying those at risk for fibrillation, designing defibrillating devices, and prescribing drugs. Winfree hoped, too, that a global, mathematical perspective on such problems would fertilize a discipline that barely existed in the United States, theoretical biology.

NOW SOME PHYSIOLOGISTS SPEAK of dynamical diseases: disorders of systems, breakdowns in coordination or control. "Systems that normally oscillate, stop oscillating, or begin to oscillate in a new and unexpected fashion, and systems that normally do not oscillate, begin oscillating," was one formulation. These syndromes include breathing disorders: panting, sighing, Cheyne-Stokes respiration, and infant apnea—linked to sudden infant death syndrome. There are dynamical blood disorders, including a form of leukemia, in which disruptions alter the balance of white and red cells, platelets and lymphocytes. Some scientists speculate that schizophrenia itself might belong in this category, along with some forms of depression.

But physiologists have also began to see chaos as health. It has long been understood that nonlinearity in feedback processes serves to regulate and control. Simply put, a linear process, given a slight nudge, tends to remain slightly off track. A nonlinear process, given the same nudge, tends to return to its starting point. Christian Huygens, the seventeenth-century Dutch physicist who helped invent both the pendulum clock and the classical science of dynamics, stumbled upon one of the great examples of this form of regulation, or so the standard story goes. Huygens noticed one day that a set of pendulum clocks placed against a wall happened to be swinging in perfect chorus-line synchronization. He knew that the clocks could not be that accurate. Nothing in the mathematical description then available for a pendulum could explain this mysterious propagation of order from one pendulum to another. Huygens surmised, correctly, that the clocks were coordinated by vibrations transmitted through the wood. This phenomenon, in which one regular cycle locks into another, is now called entrainment, or mode locking. Mode locking explains why the moon always faces the earth, or more generally why satellites tend to spin in some whole-number ratio of their orbital period: 1 to 1, or 2 to 1, or 3 to 2. When the ratio is close to a whole number, nonlinearity in the tidal attraction of the satellite tends to lock it in. Mode locking occurs throughout electronics, making it possible, for example, for a radio receiver to lock in on signals even when there are small fluctuations in their frequency. Mode locking accounts for the ability of groups of oscillators, including biological oscillators, like heart cells and nerve cells, to work in synchronization. A spectacular example in nature is a Southeast Asian species of firefly that congregates in trees during mating periods, thousands at one time, blinking in a fantastic spectral harmony.

With all such control phenomena, a critical issue is robustness: how well can a system withstand small jolts. Equally critical in biological systems is flexibility: how well can a system function over a *range* of frequencies. A locking-in to a single mode can be enslavement, preventing a system from adapting to change. Organisms must respond to circumstances that vary rapidly and unpredictably; no heartbeat or respiratory rhythm can be locked into the strict periodicities of the simplest physical models, and the same is true of the subtler rhythms of the rest of the body. Some researchers, among them Ary Goldberger of Harvard Medical School, proposed that healthy dynamics were marked by fractal physical structures, like the branching networks of bronchial tubes in the lung and conducting fibers in the heart, that allow a wide range of rhythms. Thinking of Robert Shaw's arguments, Goldberger noted: "Fractal processes associated with scaled, broadband spectra are 'information-rich.' Periodic states, in contrast, reflect narrow-band spectra and are defined by monotonous, repetitive sequences, depleted of information content." Treating such disorders, he and other physiologists suggested, may depend on broadening a system's spectral reserve, its ability to range over many different frequencies without falling into a locked periodic channel.

Arnold Mandell, the San Diego psychiatrist and dynamicist

CHAOTIC HARMONIES. The interplay of different rhythms, such as radio frequencies or planetary orbits, produces a special version of chaos. Below and on the facing page, computer pictures of some of the "attractors" that can result when three rhythms come together.

James A. Yorke

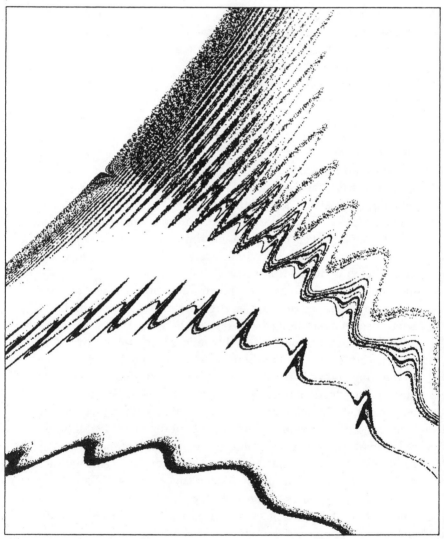

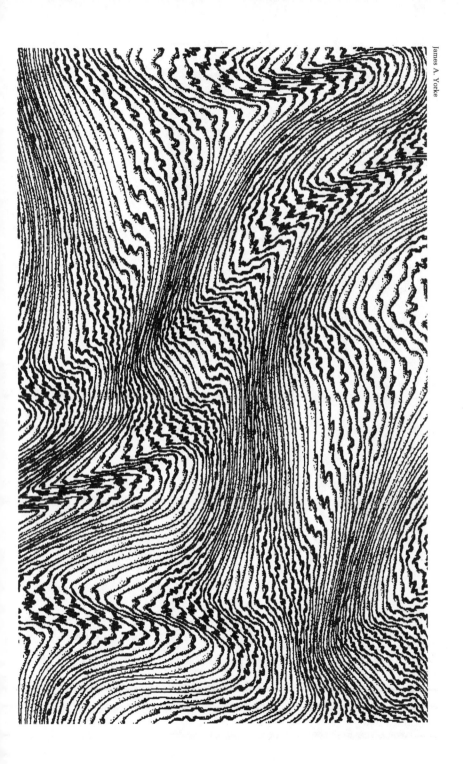

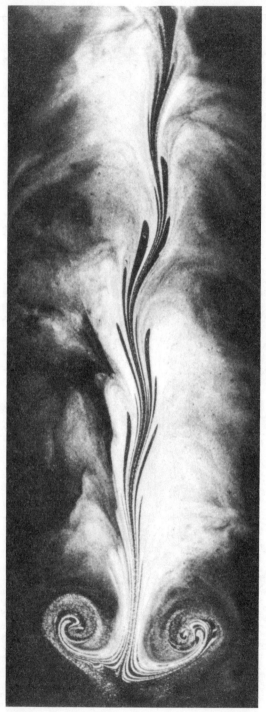

CHAOTIC FLOWS. A rod drawn through viscous fluid causes a simple, wavy form. If drawn several times, more complicated forms arise.

who came to Bernardo Huberman's defense over eye movement in schizophrenics, went even further on the role of chaos in physiology. "Is it possible that mathematical pathology, i.e. chaos, is health? And that mathematical health, which is the predictability and differentiability of this kind of a structure, is disease?" Mandell had turned to chaos as early as 1977, when he found "peculiar behavior" in certain enzymes in the brain that could only be accounted for by the new methods of nonlinear mathematics. He had encouraged the study of the oscillating three-dimensional entanglements of protein molecules in the same terms; instead of drawing static structures, he argued, biologists should understand such molecules as dynamical systems, capable of phase transitions. He was, as he said himself, a zealot, and his main interest remained the most chaotic organ of all. "When you reach an equilibrium in biology you're dead," he said. "If I ask you whether your brain is an equilibrium system, all I have to do is ask you not to think of elephants for a few minutes, and you know it isn't an equilibrium system."

To Mandell, the discoveries of chaos dictate a shift in clinical approaches to treating psychiatric disorders. By any objective measure, the modern business of "psychopharmacology"—the use of drugs to treat everything from anxiety and insomnia to schizophrenia itself—has to be judged a failure. Few patients, if any, are cured. The most violent manifestations of mental illness can be controlled, but with what long-term consequences, no one knows. Mandell offered his colleagues a chilling assessment of the most commonly used drugs. Phenothiazines, prescribed for schizophrenics, make the fundamental disorder worse. Tricyclic antidepressants "increase the rate of mood cycling, leading to long-term increases in numbers of relapsing psychopathologic episodes." And so on. Only lithium has any real medical success, Mandell said, and only for some disorders.

As he saw it, the problem was conceptual. Traditional methods for treating this "most unstable, dynamic, infinite-dimensional machine" were linear and reductionist. "The underlying paradigm remains: one gene → one peptide → one enzyme → one neurotransmitter → one receptor → one animal behavior → one clinical syndrome → one drug → one clinical rating scale. It dominates almost all research and treatment in psychopharmacology.

More than 50 transmitters, thousands of cell types, complex electromagnetic phenomenology, and continuous instability based autonomous activity at all levels, from proteins to the electroencephalogram—and still the brain is thought of as a chemical point-to-point switchboard." To someone exposed to the world of nonlinear dynamics the response could only be: How naïve. Mandell urged his colleagues to understand the flowing geometries that sustain complex systems like the mind.

Many other scientists began to apply the formalisms of chaos to research in artificial intelligence. The dynamics of systems wandering between basins of attraction, for example, appealed to those looking for a way to model symbols and memories. A physicist thinking of *ideas* as regions with fuzzy boundaries, separate yet overlapping, pulling like magnets and yet letting go, would naturally turn to the image of a phase space with "basins of attraction." Such models seemed to have the right features: points of stability mixed with instability, and regions with changeable boundaries. Their fractal structure offered the kind of infinitely self-referential quality that seems so central to the mind's ability to bloom with ideas, decisions, emotions, and all the other artifacts of consciousness. With or without chaos, serious cognitive scientists can no longer model the mind as a static structure. They recognize a hierarchy of scales, from neuron upward, providing an opportunity for the interplay of microscale and macroscale so characteristic of fluid turbulence and other complex dynamical processes.

Pattern born amid formlessness: that is biology's basic beauty and its basic mystery. Life sucks order from a sea of disorder. Erwin Schrödinger, the quantum pioneer and one of several physicists who made a nonspecialist's foray into biological speculation, put it this way forty years ago: A living organism has the "astonishing gift of concentrating a 'stream of order' on itself and thus escaping the decay into atomic chaos." To Schrödinger, as a physicist, it was plain that the structure of living matter differed from the kind of matter his colleagues studied. The building block of life—it was not yet called DNA—was an *aperiodic crystal.* "In physics we have dealt hitherto only with *periodic crystals.* To a humble physicist's mind, these are very interesting and complicated objects; they constitute one of the most fascinating and complex material structures by which inanimate nature puzzles his wits. Yet, compared with the aperiodic crystal, they are rather plain and dull." The difference was like the difference between wallpaper and tapestry, between the regular repetition of a pattern and the rich, coherent variation of an artist's creation. Physicists had learned only to understand wallpaper. It was no wonder they had managed to contribute so little to biology.

Schrödinger's view was unusual. That life was both orderly and complex was a truism; to see aperiodicity as the source of its special qualities verged on mystical. In Schrödinger's day, neither mathematics nor physics provided any genuine support for the idea. There were no tools for analyzing irregularity as a building block of life. Now those tools exist.

Chaos and Beyond

"The classification of the constituents of a chaos, nothing less here is essayed."

—HERMAN MELVILLE, *Moby-Dick*

TWO DECADES AGO Edward Lorenz was thinking about the atmosphere, Michel Hénon the stars, Robert May the balance of nature. Benoit Mandelbrot was an unknown IBM mathematician, Mitchell Feigenbaum an undergraduate at the City College of New York, Doyne Farmer a boy growing up in New Mexico. Most practicing scientists shared a set of beliefs about complexity. They held these beliefs so closely that they did not need to put them into words. Only later did it become possible to say what these beliefs were and to bring them out for examination.

Simple systems behave in simple ways. A mechanical contraption like a pendulum, a small electrical circuit, an idealized population of fish in a pond—as long as these systems could be reduced to a few perfectly understood, perfectly deterministic laws, their long-term behavior would be stable and predictable.

Complex behavior implies complex causes. A mechanical device, an electrical circuit, a wildlife population, a fluid flow, a biological organ, a particle beam, an atmospheric storm, a national economy—a system that was visibly unstable, unpredictable, or out of control must either be governed by a multitude of independent components or subject to random external influences.

Different systems behave differently. A neurobiologist who spent a career studying the chemistry of the human neuron without learning anything about memory or perception, an aircraft designer who used wind tunnels to solve aerodynamic problems without understanding the mathematics of turbulence, an economist who analyzed the psychology of purchasing decisions without gaining an ability to forecast large-scale trends—scientists like these, knowing that the components of their disciplines were different, took it for granted that the complex systems made up of billions of these components must also be different.

Now all that has changed. In the intervening twenty years, physicists, mathematicians, biologists, and astronomers have created an alternative set of ideas. Simple systems give rise to complex behavior. Complex systems give rise to simple behavior. And most important, the laws of complexity hold universally, caring not at all for the details of a system's constituent atoms.

For the mass of practicing scientists—particle physicists or neurologists or even mathematicians—the change did not matter immediately. They continued to work on research problems within their disciplines. But they were aware of something called chaos. They knew that some complex phenomena had been explained, and they knew that other phenomena suddenly seemed to need new explanations. A scientist studying chemical reactions in a laboratory or tracking insect populations in a three-year field experiment or modeling ocean temperature variations could not respond in the traditional way to the presence of unexpected fluctuations or oscillations—that is, by ignoring them. For some, that meant trouble. On the other hand, pragmatically, they knew that money was available from the federal government and from corporate research facilities for this faintly mathematical kind of science. More and more of them realized that chaos offered a fresh way to proceed with old data, forgotten in desk drawers because they had proved too erratic. More and more felt the compartmentalization of science as an impediment to their work. More and more felt the futility of studying parts in isolation from the whole. For them, chaos was the end of the reductionist program in science.

Uncomprehension; resistance; anger; acceptance. Those who had promoted chaos longest saw all of these. Joseph Ford of the Georgia Institute of Technology remembered lecturing to a thermodynamics group in the 1970s and mentioning that there was a chaotic behavior in the Duffing equation, a well-known textbook model for a simple oscillator subject to friction. To Ford, the presence of chaos in the Duffing equation was a curious fact—just one of those things he knew to be true, although several years passed before it was published in *Physical Review Letters*. But he might as well have told a gathering of paleontologists that dinosaurs had feathers. They knew better.

"When I said that? Jee-sus Christ, the audience began to bounce up and down. It was, 'My daddy played with the Duffing equation, and my granddaddy played with the Duffing equation, and nobody seen anything like what you're talking about.' You would really run across resistance to the notion that nature is complicated. What I didn't understand was the hostility."

Comfortable in his Atlanta office, the winter sun setting outside, Ford sipped soda from an oversized mug with the word CHAOS painted in bright colors. His younger colleague Ronald Fox talked about his own conversion, soon after buying an Apple II computer for his son, at a time when no self-respecting physicist would buy such a thing for his work. Fox heard that Mitchell Feigenbaum had discovered universal laws guiding the behavior of feedback functions, and he decided to write a short program that would let him see the behavior on the Apple display. He saw it all painted across the screen—pitchfork bifurcations, stable lines breaking in two, then four, then eight; the appearance of chaos itself; and within the chaos, the astonishing geometric regularity. "In a couple of days you could redo all of Feigenbaum," Fox said. Self-teaching by computing persuaded him and others who might have doubted a written argument.

Some scientists played with such programs for a while and then stopped. Others could not help but be changed. Fox was one of those who had remained conscious of the limits of standard linear science. He knew he had habitually set the hard nonlinear problems aside. In practice a physicist would always end up saying, *This is a problem that's going to take me to the handbook of special functions, which is the last place I want to go, and I'm sure as hell not going to get on a machine and do it, I'm too sophisticated for that.*

"The general picture of nonlinearity got a lot of people's attention—slowly at first, but increasingly," Fox said. "Everybody that looked at it, it bore fruit for. You now look at any problem you looked at before, no matter what science you're in. There was a place where you quit looking at it because it became nonlinear. Now you know how to look at it and you go back."

Ford said, "If an area begins to grow, it has to be because some clump of people feel that there's something it offers them— that if they modify their research, the rewards could be very big. To me chaos is like a dream. It offers the possibility that, if you come over and play this game, you can strike the mother lode."

Still, no one could quite agree on the word itself.

Philip Holmes, a white-bearded mathematician and poet from Cornell by way of Oxford: *The complicated, aperiodic, attracting orbits of certain (usually low-dimensional) dynamical systems.*

Hao Bai-Lin, a physicist in China who assembled many of the historical papers of chaos into a single reference volume: *A kind of order without periodicity.* And: *A rapidly expanding field of research to which mathematicians, physicists, hydrodynamicists, ecologists and many others have all made important contributions.* And: *A newly recognized and ubiquitous class of natural phenomena.*

H. Bruce Stewart, an applied mathematician at Brookhaven National Laboratory on Long Island: *Apparently random recurrent behavior in a simple deterministic (clockwork-like) system.*

Roderick V. Jensen of Yale University, a theoretical physicist exploring the possibility of quantum chaos: *The irregular, unpredictable behavior of deterministic, nonlinear dynamical systems.*

James Crutchfield of the Santa Cruz collective: *Dynamics with positive, but finite, metric entropy. The translation from mathese is: behavior that produces information (amplifies small uncertainties), but is not utterly unpredictable.*

And Ford, self-proclaimed evangelist of chaos: *Dynamics freed at last from the shackles of order and predictability. . . . Systems liberated to randomly explore their every dynamical possibility. . . . Exciting variety, richness of choice, a cornucopia of opportunity.*

John Hubbard, exploring iterated functions and the infinite fractal wildness of the Mandelbrot set, considered chaos a poor name for his work, because it implied randomness. To him, the overriding message was that simple processes in nature could produce magnificent edifices of complexity *without* randomness.

In nonlinearity and feedback lay all the necessary tools for encoding and then unfolding structures as rich as the human brain.

To other scientists, like Arthur Winfree, exploring the global topology of biological systems, chaos was too narrow a name. It implied simple systems, the one-dimensional maps of Feigenbaum and the two- or three- (and a fraction) dimensional strange attractors of Ruelle. Low-dimensional chaos was a special case, Winfree felt. He was interested in the laws of many-dimensional complexity—and he was convinced that such laws existed. Too much of the universe seemed beyond the reach of low-dimensional chaos.

The journal *Nature* carried a running debate about whether the earth's climate followed a strange attractor. Economists looked for recognizable strange attractors in stock market trends but so far had not found them. Dynamicists hoped to use the tools of chaos to explain fully developed turbulence. Albert Libchaber, now at the University of Chicago, was turning his elegant experimental style to the service of turbulence, creating a liquid-helium box thousands of times larger than his tiny cell of 1977. Whether such experiments, liberating fluid disorder in both space and time, would find simple attractors, no one knew. As the physicist Bernardo Huberman said, "If you had a turbulent river and put a probe in it and said, 'Look, here's a low-dimensional strange attractor,' we would all take off our hats and look."

Chaos was the set of ideas persuading all these scientists that they were members of a shared enterprise. Physicist or biologist or mathematician, they believed that simple, deterministic systems could breed complexity; that systems too complex for traditional mathematics could yet obey simple laws; and that, whatever their particular field, their task was to understand complexity itself.

"LET US AGAIN LOOK at the laws of thermodynamics," wrote James E. Lovelock, author of the Gaia hypothesis. "It is true that at first sight they read like the notice at the gate of Dante's Hell . . ." But.

The Second Law is one piece of technical bad news from science that has established itself firmly in the nonscientific culture. Everything tends toward disorder. Any process that converts energy from one form to another must lose some as heat. Perfect efficiency is impossible. The universe is a one-way street. *Entropy must always increase in the universe and in any hypothetical isolated system within it.* However expressed, the Second Law is a rule from which there seems no appeal. In thermodynamics that is true. But the Second Law has had a life of its own in intellectual realms far removed from science, taking the blame for disintegration of societies, economic decay, the breakdown of manners, and many other variations on the decadent theme. These secondary, metaphorical incarnations of the Second Law now seem especially misguided. In our world, complexity flourishes, and those looking to science for a general understanding of nature's habits will be better served by the laws of chaos.

Somehow, after all, as the universe ebbs toward its final equilibrium in the featureless heat bath of maximum entropy, it manages to create interesting structures. Thoughtful physicists concerned with the workings of thermodynamics realize how disturbing is the question of, as one put it, "how a purposeless flow of energy can wash life and consciousness into the world." Compounding the trouble is the slippery notion of entropy, reasonably well-defined for thermodynamic purposes in terms of heat and temperature, but devilishly hard to pin down as a measure of *disorder*. Physicists have trouble enough measuring the degree of order in water, forming crystalline structures in the transition to ice, energy bleeding away all the while. But thermodynamic entropy fails miserably as a measure of the changing degree of form and formlessness in the creation of amino acids, of microorganisms, of self-reproducing plants and animals, of complex information systems like the brain. Certainly these evolving islands of order must obey the Second Law. The important laws, the creative laws, lie elsewhere.

Nature forms patterns. Some are orderly in space but disorderly in time, others orderly in time but disorderly in space. Some patterns are fractal, exhibiting structures self-similar in scale. Others give rise to steady states or oscillating ones. Pattern formation has become a branch of physics and of materials science, allowing scientists to model the aggregation of particles into clusters, the fractured spread of electrical discharges, and the growth of crystals in ice and metal alloys. The dynamics seem so basic—shapes changing in space and time—yet only now are the tools available to understand them. It is a fair question now to ask a physicist, "Why are all snowflakes different?"

Ice crystals form in the turbulent air with a famous blending of symmetry and chance, the special beauty of six-fold indeterminacy. As water freezes, crystals send out tips; the tips grow, their boundaries becoming unstable, and new tips shoot out from the sides. Snowflakes obey mathematical laws of surprising subtlety, and it was impossible to predict precisely how fast a tip would grow, how narrow it would be, or how often it would branch. Generations of scientists sketched and cataloged the variegated patterns: plates and columns, crystals and polycrystals, needles and dendrites. The treatises treated crystal formation as a classification matter, for lack of a better approach.

Growth of such tips, dendrites, is now known as a highly nonlinear unstable free boundary problem, meaning that models need to track a complex, wiggly boundary as it changes dynamically. When solidification proceeds from outside to inside, as in an ice tray, the boundary generally remains stable and smooth, its speed controlled by the ability of the walls to draw away the heat. But when a crystal solidifies outward from an initial seed—as a snowflake does, grabbing water molecules while it falls through the moisture-laden air—the process becomes unstable. Any bit of boundary that gets out ahead of its neighbors gains an advantage in picking up new water molecules and therefore grows that much faster—the "lightning-rod effect." New branches form, and then subbranches.

One difficulty was in deciding which of the many physical forces involved are important and which can safely be ignored. Most important, as scientists have long known, is the diffusion of the heat released when water freezes. But the physics of heat diffusion cannot completely explain the patterns researchers observe when they look at snowflakes under microscopes or grow them in the laboratory. Recently scientists worked out a way to incorporate another process: surface tension. The heart of the new snowflake model is the essence of chaos: a delicate balance between forces of stability and forces of instability; a powerful interplay of forces on atomic scales and forces on everyday scales.

Oscar Kapp, inset: Shoudon Liang

Where heat diffusion tends to create instability, surface tension creates stability. The pull of surface tension makes a substance prefer smooth boundaries like the wall of a soap bubble. It costs energy to make surfaces that are rough. The balancing of these tendencies depends on the size of the crystal. While diffusion is mainly a large-scale, macroscopic process, surface tension is strongest at the microscopic scales.

Traditionally, because the surface tension effects are so small, researchers assumed that for practical purposes they could disregard them. Not so. The tiniest scales proved crucial; there the surface effects proved infinitely sensitive to the molecular structure of a solidifying substance. In the case of ice, a natural molecular symmetry gives a built-in preference for six directions of growth. To their surprise, scientists found that the mixture of stability and instability manages to amplify this microscopic preference, creating the near-fractal lacework that makes snowflakes. The mathematics came not from atmospheric scientists but from theoretical physicists, along with metallurgists, who had their own interest. In metals the molecular symmetry is different, and so are the characteristic crystals, which help determine an alloy's strength. But the mathematics are the same: the laws of pattern formation are universal.

Sensitive dependence on initial conditions serves not to destroy but to create. As a growing snowflake falls to earth, typically floating in the wind for an hour or more, the choices made by the branching tips at any instant depend sensitively on such things as the temperature, the humidity, and the presence of impurities in the atmosphere. The six tips of a single snowflake, spreading within a millimeter space, feel the same temperatures, and because the laws of growth are purely deterministic, they maintain a near-perfect symmetry. But the nature of turbulent air is such that any pair of snowflakes will experience very different paths. The final flake records the history of all the changing weather conditions it has experienced, and the combinations may as well be infinite.

BRANCHING AND CLUMPING, (*on facing page*). The study of pattern formation, encouraged by fractal mathematics, brought together such natural patterns as the lightning-like paths of an electrical discharge and the simulated aggregation of randomly moving particles (*inset*).

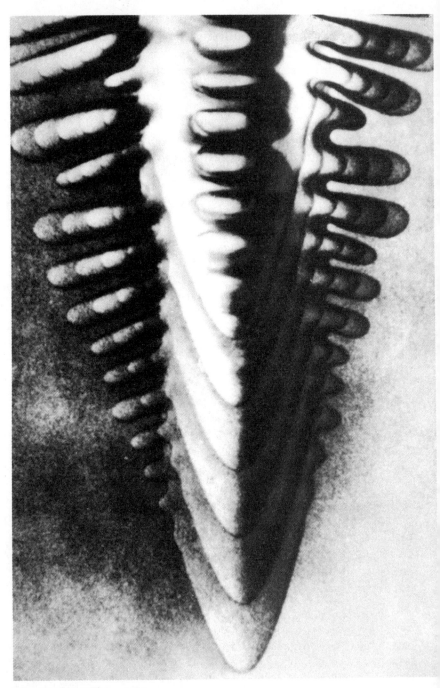

Martin Glicksman / Fereydoon Family

BALANCING STABILITY AND INSTABILITY. As a liquid crystallizes, it forms a growing tip (shown in a multiple-exposure photograph) with a boundary that becomes unstable and sends off side-branches (*left*). Computer simulations of the delicate thermodynamic processes mimic real snowflakes (*above*).

Snowflakes are nonequilibrium phenomena, physicists like to say. They are products of imbalance in the flow of energy from one piece of nature to another. The flow turns a boundary into a tip, the tip into an array of branches, the array into a complex structure never before seen. As scientists have discovered such instability obeying the universal laws of chaos, they have succeeded in applying the same methods to a host of physical and chemical problems, and, inevitably, they suspect that biology is next. In the back of their minds, as they look at computer simulations of dendrite growth, they see algae, cell walls, organisms budding and dividing.

From microscopic particles to everyday complexity, many paths now seem open. In mathematical physics the bifurcation theory of Feigenbaum and his colleagues advances in the United States and Europe. In the abstract reaches of theoretical physics scientists probe other new issues, such as the unsettled question of quantum chaos: Does quantum mechanics admit the chaotic phenomena of classical mechanics? In the study of moving fluids Libchaber builds his giant liquid-helium box, while Pierre Hohenberg and Günter Ahlers study the odd-shaped traveling waves of convection. In astronomy chaos experts use unexpected gravitational instabilities to explain the origin of meteorites—the seemingly inexplicable catapulting of asteroids from far beyond Mars. Scientists use the physics of dynamical systems to study the human immune system, with its billions of components and its capacity for learning, memory, and pattern recognition, and they simultaneously study evolution, hoping to find universal mechanisms of adaptation. Those who make such models quickly see structures that replicate themselves, compete, and evolve by natural selection.

"Evolution is chaos with feedback," Joseph Ford said. The universe is randomness and dissipation, yes. But randomness with direction can produce surprising complexity. And as Lorenz discovered so long ago, dissipation is an agent of order.

"God plays dice with the universe," is Ford's answer to Einstein's famous question. "But they're loaded dice. And the main objective of physics now is to find out by what rules were they loaded and how can we use them for our own ends."

SUCH IDEAS HELP drive the collective enterprise of science forward. Still, no philosophy, no proof, no experiment ever seems quite enough to sway the individual researchers for whom science must first and always provide a way of working. In some laboratories, the traditional ways falter. Normal science goes astray, as Kuhn put it; a piece of equipment fails to meet expectations; "the profession can no longer evade anomalies." For any one scientist the ideas of chaos could not prevail until the method of chaos became a necessity.

Every field had its own examples. In ecology, there was William M. Schaffer, who trained as the last student of Robert MacArthur, the dean of the field in the fifties and sixties. MacArthur built a conception of nature that gave a firm footing to the idea of *natural balance*. His models supposed that equilibriums would exist and that populations of plants and animals would remain close to them. To MacArthur, balance in nature had what could almost be called a moral quality—states of equilibrium in his models entailed the most efficient use of food resources, the least waste. Nature, if left alone, would be good.

Two decades later MacArthur's last student found himself realizing that ecology based on a sense of equilibrium seems doomed to fail. The traditional models are betrayed by their linear bias. Nature is more complicated. Instead he sees chaos, "both exhilarating and a bit threatening." Chaos may undermine ecology's most enduring assumptions, he tells his colleagues. "What passes for fundamental concepts in ecology is as mist before the fury of the storm—in this case, a full, nonlinear storm."

Schaffer is using strange attractors to explore the epidemiology of childhood diseases such as measles and chicken pox. He has collected data, first from New York City and Baltimore, then from Aberdeen, Scotland, and all England and Wales. He has made a dynamical model, resembling a damped, driven pendulum. The diseases are driven each year by the infectious spread among children returning to school, and damped by natural resistance. Schaffer's model predicts strikingly different behavior for these diseases. Chicken pox should vary periodically. Measles should vary chaotically. As it happens, the data show exactly what Schaffer predicts. To a traditional epidemiologist the yearly variations in measles seemed inexplicable—random and noisy. Schaffer, using the techniques of phase-space reconstruction, shows that measles follow a strange attractor, with a fractal dimension of about 2.5.

Schaffer computed Lyapunov exponents and made Poincaré maps. "More to the point," Schaffer said, "if you look at the pictures it jumps out at you, and you say, 'My God, this is the same thing.' " Although the attractor is chaotic, some predictability becomes possible in light of the deterministic nature of the model. A year of high measles infection will be followed by a crash. After a year of medium infection, the level will change only slightly. A year of low infection produces the greatest unpredictability. Schaffer's model also predicted the consequences of damping the dynamics by mass inoculation programs—consequences that could not be predicted by standard epidemiology.

On the collective scale and on the personal scale, the ideas of chaos advance in different ways and for different reasons. For Schaffer, as for many others, the transition from traditional science to chaos came unexpectedly. He was a perfect target for Robert May's evangelical plea in 1975; yet he read May's paper and discarded it. He thought the mathematical ideas were unrealistic for the kinds of systems a practicing ecologist would study. Oddly, he knew too much about ecology to appreciate May's point. These were one-dimensional maps, he thought—what bearing could they have on continuously changing systems? So a colleague said, "Read Lorenz." He wrote the reference on a slip of paper and never bothered to pursue it.

Years later Schaffer lived in the desert outside of Tucson, Arizona, and summers found him in the Santa Catalina mountains just to the north, islands of chaparral, merely hot when the desert floor is roasting. Amid the thickets in June and July, after the spring blooming season and before the summer rain, Schaffer and his graduate students tracked bees and flowers of different species. This ecological system was easy to measure despite all its year-to-year variation. Schaffer counted the bees on every stalk, measured the pollen by draining flowers with pipettes, and analyzed the data mathematically. Bumblebees competed with honeybees,

and honeybees competed with carpenter bees, and Schaffer made
a convincing model to explain the fluctuations in population.
By 1980 he knew that something was wrong. His model broke
down. As it happened, the key player was a species he had overlooked: ants. Some colleagues suspected unusual winter weather;
others unusual summer weather. Schaffer considered complicat-
ing his model by adding more variables. But he was deeply frustrated. Word was out among the graduate students that summer
at 5,000 feet with Schaffer was hard work. And then everything
changed.

He happened upon a preprint about chemical chaos in a complicated laboratory experiment, and he felt that the authors had
experienced exactly his problem: the impossibility of monitoring
dozens of fluctuating reaction products in a vessel matched the
impossibility of monitoring dozens of species in the Arizona
mountains. Yet they had succeeded where he had failed. He read
about reconstructing phase space. He finally read Lorenz, and
Yorke, and others. The University of Arizona sponsored a lecture
series on "Order in Chaos." Harry Swinney came, and Swinney
knew how to talk about experiments. When he explained chemical
chaos, displaying a transparency of a strange attractor, and said,
"That's real data," a chill ran up Schaffer's spine.

"All of a sudden I knew that that was my destiny," Schaffer
said. He had a sabbatical year coming. He withdrew his application for National Science Foundation money and applied for a
Guggenheim Fellowship. Up in the mountains, he knew, the ants
changed with the season. Bees hovered and darted in a dynamical
buzz. Clouds skidded across the sky. He could not work the old
way any more.

Afterword

EVEN NOW, *CHAOS THEORY* sounds like a bit of an oxymoron. In the 1980s, "chaos" and "theory" were words that didn't seem to belong in the same room, let alone the same sentence. When friends heard that I was researching a book about chaos—and that it was to do with science—there were quizzical looks and raised eyebrows. Much later, one told me she had thought I was writing about "gas." As it says in the subtitle, chaos was a *new* science—strange and alien-sounding, exciting and hard to accept.

What a difference twenty years make. The ideas of chaos have been adopted and internalized, not just by mainstream science but also by the culture at large. Still, even now, plenty of scientists find chaos to be strange and alien-sounding, exciting and hard to accept.

We've all now heard of chaos, at least a little. "I'm still not clear on chaos," says Laura Dern's character in the 1993 film *Jurassic Park,* so that Jeff Goldblum's character—who announces himself as a "chaotician"—can explain flirtatiously, "It simply deals with unpredictability in complex systems. . . . A butterfly can flap its wings in Peking, and in Central Park you get rain instead of sunshine." By then the Butterfly Effect was well on its way to becoming a pop-culture cliché: inspiring at least two movies, an entry in *Bartlett's Quotations,* a music video, and a thousand Web sites and blogs. (Only the place names keep changing: the butterfly flaps its wings in Brazil, Peru, China, California, Tahiti, and South America, and the rain/hurricane/tornado/storm arrives in Texas, Florida, New York, Nebraska, Kansas, and Central Park.) After the big hurricanes of 2006, *Physics Today* published an article titled "Battling the Butterfly Effect," whimsically blaming butterflies in battalions: "Visions of *Lepidoptera* terrorist training camps spring suddenly to mind."

Aspects of chaos—different aspects, usually—have been taken up by modern management theorists on the one hand, and postmodern literary theorists on the other. Both camps have found use for phrases like "orderly disorder," especially popular in dissertation titles. Compelling literary characters, such as Shakespeare's Cleopatra, are seen to be "strange attractors." So are chart patterns in the financial markets. Meanwhile, painters as well as sculptors have found inspiration in both the words and the images of fractal geometry. For my money, the most powerful artistic incarnation of these ideas came in Tom Stoppard's play *Arcadia,* which opened in London a few months before *Jurassic Park.* It, too, features a mathematician reveling in chaos: "The freaky stuff," he says, "is turning out to be the mathematics of the natural world." Stoppard goes beyond orderly disorder to the tension between the formal English garden and the wilderness, between the classical and the Romantic. He is channeling the voices in this book, and to quote him here is to engage in loopy feedback, but I can't help it. He captures the exhilaration of so many young researchers at the discovery of chaos. He sees the opening door and the vista beyond.

> *The ordinary-sized stuff which is our lives, the things people write poetry about—clouds—daffodils—waterfalls—and what happens in a cup of coffee when the cream goes in—these things are full of mystery, as mysterious to us as the heavens were to the Greeks. . . . The future is disorder. A door like this has cracked open five or six times since we got up on our hind legs. It's the best possible time to be alive, when almost everything you thought you knew was wrong.*

The door is open more than a crack now, and a new generation of scientists has come along, armed with a more robust set of assumptions about how nature works. They know that a complex dynamical system can get freaky. They know, when it does that, that you can still look it in the eye and take its measure. Meetings across disciplinary lines to share methodologies on scaling patterns or network behaviors are now, if not the rule, at least no longer the exception.

By and large, the pioneers of chaos came in from the wilderness and took their places in the scientific establishment. Edward Lorenz, as a much-honored professor emeritus at M.I.T., was still seen coming to work in his nineties and watching the weather from his office high up in Building 54. Mitchell Feigenbaum joined Rockefeller University and created a mathematical physics laboratory there. Robert May became president of the Royal Society and chief scientific adviser to the government of the U.K. and, in 2001, was created Baron May of Oxford. As for Benoit Mandelbrot, a "Vita" he published on his Yale Web page in 2006 listed twenty-four awards, prizes, and medals, two decorations, nineteen "diplomas, *honoris causa* & the like," twelve memberships in scientific societies, fifteen memberships on editorial boards and committees, and a variety of items bearing his name, including a "Tree along the Nobel Lane" in Balantonfüred, Hungary, a laboratory in China, and an asteroid.

The principles they discovered and the concepts they invented have continued to evolve—beginning with the word "chaos" itself. Already by the mid-1980s the word was being defined rather narrowly (see page 306) by many scientists, who applied it to a special subset of the phenomena covered by more general terms such as "complex systems." Astute readers, though, could tell that I preferred Joe Ford's more freewheeling "cornucopia" style of definition—"Dynamics freed at last from the shackles of order and predictability . . ."—and still do. But everything evolves in the direction of specialization, and strictly speaking, "chaos" is now a very particular thing. When Yaneer Bar-Yam wrote a kilopage textbook, *Dynamics of Complex Systems,* in 2003, he took care of chaos proper in the first section of the first chapter. ("The first chapter, I have to admit, is 300 pages, okay?" he says.) Then came Stochastic Processes, Modeling Simulation, Cellular Automata, Computation Theory and Information Theory, Scaling, Renormalization, and

Fractals, Neural Networks, Attractor Networks, Homogenous Systems, Inhomogenous Systems, and so on.

Bar-Yam, the son of a high-energy physicist, had studied condensed matter physics and become an engineering professor at Boston University, but he left in 1997 to found the New England Complex Systems Institute. He had been exposed to Stephen Wolfram's work on cellular automata and Robert Devaney's work in chaos and discovered that he was less interested in polymers and superconductors than in neural networks and—he says this with no sense of grandiosity—the nature of human civilization. "Thinking about civilization," he says, "led me to think about complexity as an entity. How do you compare civilization to something else? Is it like brass? Is it like a frog? How do you answer that question? This is what motivates complex systems."

In case you couldn't tell, civilization is more like a frog than brass. For one thing, it evolves—evolutionary, adaptive processes being essential in the design and creation of anything so complex that it cannot effectively be decomposed into separate pieces. Socioeconomic systems are like ecosystems. In fact, they *are* ecosystems. With computer modeling, Bar-Yam has been studying, among other things, global patterns of ethnic violence, trying to isolate patterns of population mixing and boundaries that trigger conflicts. At its core, this is research on pattern formation. That he can do this work at all illustrates the profound shift over the past two decades in the community's understanding of what constitutes a legitimate scientific problem. "Let me diagram for you the process," he says. He has a parable:

People are working to harvest fruit from an orchard, okay? Beautiful fruit were taken and brought to market, and then you harvest fruit that's higher up in the trees. It's a little bit harder to get to and maybe a little bit smaller and not as nice. And then you build ladders and you climb up the tree and you get to the higher fruit. And then you reward people for building the ladders.

My feeling of what I did is, I looked and I saw that there was a hedge, and beyond the hedge was another orchard, which had beautiful fruit on many, many trees. And here I am, I find

*a fruit and I go back through the hedge and I show it to peo-
ple. And they say, "That's not a fruit!" They couldn't recognize the fruit anymore.*

Communication is better now, he feels. Disciplines across the scientific spectrum have learned to focus on understanding complexity and scale and patterns and the collective behavior that is
associated with patterns. That's fruit.

IN THE HEADY early days, researchers described chaos as the
century's third revolution in the physical sciences, after relativity
and quantum mechanics. What has become clear now is that chaos
is *inextricable* from relativity and quantum mechanics. There is
only one physics.

The fundamental equations of general relativity are nonlinear—already a signal, we know by now, that chaos lurks. "People aren't always well versed in its methods," says Janna Levin, an
astrophysicist and cosmologist at Barnard College of Columbia
University. "Theoretical physics in particular is built on the notion
of fundamental symmetries," she notes. "For that reason, I think
it's been a difficult paradigm shift for theoretical physics to embrace." Symmetries and symmetry groups tend to produce solvable equations—that's why they work so well. When they work.

As a relativist, Levin deals in the biggest questions there are.
(Is the universe infinite, for example, or just really big? Her work
suggests *big*, or—if we want to be technical—topologically compact and multiconnected.) In studying the origin of the universe,
Levin found herself dealing with chaos willy-nilly and ran into
resistance. "When I first brought this work out, there was an in-
sanely violent reaction against it," she says. People thought chaos
was fine "for complicated, grungy physical systems—not the pure,
uncomplicated and virtual terrain of fundamental physics."

*We were working on chaos in pure general relativity without
any grunge, and this was a tiny, tiny, little industry—working
out chaos in a generic big bang, or collapse to a black hole, or
in orbits around a black hole. People don't think it's a spooky*

word, but they're surprised to see chaos play a role in some-thing as ungrungy—no atoms or junk—as a purely relativistic system.

Astronomers had already found the fingerprints of chaos in vio-lence on the sun's surface, gaps in the asteroid belt, and the distri-bution of galaxies. Levin and her colleagues have found them in the exit from the big bang and in black holes. They predict that light trapped by a black hole can enter unstable chaotic orbits and be reemitted—making the black hole visible, if only briefly. Yes, chaos can light up black holes. "There are rational numbers to mine, fractal sets, and all kinds of truly beautiful consequences," she says. "So on the one hand, people are horrified, on the other they're mesmerized." She does chaos in curved space-time. Ein-stein would be proud.

AS FOR ME, I never returned to chaos, but readers might spot seeds of all my later books in this one. I knew hardly anything about Richard Feynman, but he has a cameo here (see page 137). Isaac Newton has more than a cameo: he seems to be the antihero of chaos, or the god to be overthrown. I discovered only later, read-ing his notebooks and letters, how wrong I'd been about him. And for twenty years I've been pursuing a thread that began with some-thing Rob Shaw told me, about chaos and information theory, as invented by Claude Shannon. Chaos is a creator of *information*—another apparent paradox. This thread connects with something Bernardo Huberman said: that he was seeing complex behaviors emerge unexpectedly in information networks. Something was dawning, and we're finally starting to see what it is.

James Gleick
Key West
February 2008

Notes on Sources
and Further Reading

THIS BOOK DRAWS on the words of about two hundred scientists, in public lectures, in technical writing, and most of all in interviews conducted from April 1984 to December 1986. Some of the scientists were specialists in chaos; others were not. Some made themselves available for many hours over a period of months, offering insights into the history and practice of science that are impossible to credit fully. A few provided unpublished written recollections.

Few useful secondary sources of information on chaos exist, and the lay reader in search of further reading will find few places to turn. Perhaps the first general introduction to chaos—still eloquently conveying the flavor of the subject and outlining some of the fundamental mathematics—was Douglas R. Hofstadter's November 1981 column in *Scientific American*, reprinted in *Metamagical Themas* (New York: Basic Books, 1985). Two useful collections of the most influential scientific papers are Hao Bai-Lin, *Chaos* (Singapore: World Scientific, 1984) and Predrag Cvitanović, *Universality in Chaos* (Bristol: Adam Hilger, 1984). Their selections overlap surprisingly little; the former is perhaps a bit more historically oriented. For anyone interested in the origins of fractal geometry, the indispensable, encyclopedic, exasperating source is Benoit Mandelbrot, *The Fractal Geometry of Nature* (New York: Freeman, 1977). *The Beauty of Fractals,* Heinz-Otto Peitgen and Peter H. Richter (Berlin: Springer-Verlag, 1986), delves into many areas of the mathematics of chaos in European-Romantic fashion, with invaluable essays by Mandelbrot, Adrien Douady, and Gert Eilenberger; it contains many lavish color and black-and-white graphics, several of which are reproduced in this book. A well-illustrated text directed at engineers and others seeking a practical survey of the mathematical ideas is H. Bruce Stewart and J. M. Thompson, *Nonlinear Dynamics and Chaos* (Chichester: Wiley, 1986). None of these books will be valuable to readers without some technical background.

In describing the events of this book and the motivations and perspectives of the scientists, I have avoided the language of science wherever possible, assuming that the technically aware will know when they are reading about integrability, power-law distribution, or complex analysis. Readers who want mathematical elaboration or specific references will find them in the chapter notes below. In selecting a few journal articles from the thousands that might have been cited, I chose either those which most directly influenced the events chronicled in this book or those which will be most broadly useful to readers seeking further context for ideas that interest them.

Descriptions of places are generally based on my visits to the sites. The following institutions made available their researchers, their libraries, and in some cases their computer facilities: Boston University, Cornell University, Courant Institute of Mathematics, European Centre for Medium Range Weather Forecasts, Georgia Institute of Technology, Harvard University, IBM Thomas J. Watson Research Center, Institute for Advanced Study, Lamont-Doherty Geophysical Observatory, Los Alamos National Laboratory, Massachusetts Institute of Technology, National Center for Atmospheric Research, National Institutes of Health, National Meteorological Center, New York University, Observatoire de Nice, Princeton University, University of California at Berkeley, University of California at Santa Cruz, University of Chicago, Woods Hole Oceanographic Institute, Xerox Palo Alto Research Center.

For particular quotations and ideas, the notes below indicate my principal sources. I give full citations for books and articles; where only a last name is cited, the reference is to one of the following scientists, who were especially helpful to my research:

Günter Ahlers	Ralph E. Gomory
F. Tito Arecchi	Stephen Jay Gould
Michael Barnsley	John Guckenheimer
Lennart Bengtsson	Brosl Hasslacher
William D. Bonner	Michel Hénon
Robert Buchal	Douglas R. Hofstadter
William Burke	Pierre Hohenberg
David Campbell	Frank Hoppensteadt
Peter A. Carruthers	Hendrik Houthakker
Richard J. Cohen	John H. Hubbard
James Crutchfield	Bernardo Huberman
Predrag Cvitanović	Raymond E. Ideker
Minh Duong-van	Erica Jen
Freeman Dyson	Roderick V. Jensen
Jean-Pierre Eckmann	Leo Kadanoff
Fereydoon Family	Donald Kerr
J. Doyne Farmer	Joseph Klafter
Mitchell J. Feigenbaum	Thomas S. Kuhn
Joseph Ford	Mark Laff
Ronald Fox	Oscar Lanford
Robert Gilmore	James Langer
Leon Glass	Joel Lebowitz
James Glimm	Cecil E. Leith
Ary L. Goldberger	Herbert Levine
Jerry P. Gollub	Albert Libchaber

Edward N. Lorenz
Willem Malkus
Benoit Mandelbrot
Arnold Mandell
Syukuro Manabe
Arnold J. Mandell
Philip Marcus
Paul C. Martin
Robert M. May
Francis C. Moon
Jürgen Moser
David Mumford
Michael Nauenberg
Norman Packard
Heinz-Otto Peitgen
Charles S. Peskin
James Ramsey
Peter H. Richter
Otto Rössler
David Ruelle
William M. Schaffer
Stephen H. Schneider
Christopher Scholz

Robert Shaw
Michael F. Shlesinger
Yasha G. Sinai
Steven Smale
Edward A. Spiegel
H. Bruce Stewart
Steven Strogatz
Harry Swinney
Tomas Toffoli
Felix Villars
William M. Visscher
Richard Voss
Bruce J. West
Robert White
Gareth P. Williams
Kenneth G. Wilson
Arthur T. Winfree
Jack Wisdom
Helena Wisniewski
Steven Wolfram
J. Austin Woods
James A. Yorke

PROLOGUE

1   Los Alamos   Feigenbaum, Carruthers, Campbell, Farmer, Visscher, Kerr, Hasslacher, Jen.
2   "I understand you're"   Feigenbaum, Carruthers.
4   Government program   Buchal, Shlesinger, Wisniewski.
5   elements of motion   Yorke.
5   process rather than state   F. K. Browand, "The Structure of the Turbulent Mixing Layer," *Physica* 18D (1986), p. 135.
5   the behavior of cars   Japanese scientists took the traffic problem especially seriously; e.g., Toshimitsu Musha and Hideyo Higuchi, "The 1/f Fluctuation of a Traffic Current on an Expressway," *Japanese Journal of Applied Physics* (1976), pp. 1271–75.
5   That realization   Mandelbrot, Ramsey; Wisdom, Marcus; Alvin M. Saperstein, "Chaos—A Model for the Outbreak of War," *Nature* 309 (1984), pp. 303–5.
5   "Fifteen years ago   Shlesinger.
6   just three things   Shlesinger.
6   third great revolution   Ford.
6   "Relativity eliminated"   Joseph Ford, "What Is Chaos, That We Should Be Mindful of It?" preprint, Georgia Institute of Technology, p. 12.
6   The cosmologist   John Boslough, *Stephen Hawking's Universe* (Cambridge: Cambridge University Press, 1980); see also Robert Shaw, *The Dripping Faucet as a Model Chaotic System* (Santa Cruz: Aerial, 1984), p. 1.

THE BUTTERFLY EFFECT

11 THE SIMULATED WEATHER   Lorenz, Malkus, Spiegel, Farmer. The essential Lorenz is a triptych of papers whose centerpiece is "Deterministic Nonperiodic Flow," *Journal of the Atmospheric Sciences* 20 (1963), pp. 130–41; flanking this are "The Mechanics of Vacillation," *Journal of the Atmospheric Sciences* 20 (1963), pp. 448–64, and "The Problem of Deducing the Climate from the Governing Equations," *Tellus* 16 (1964), pp. 1–11. They form a deceptively elegant piece of work that continues to influence mathematicians and physicists twenty years later. Some of Lorenz's personal recollections of his first computer model of the atmosphere appear in "On the Prevalence of Aperiodicity in Simple Systems," in *Global Analysis*, eds. Mgrmela and J. Marsden (New York: Springer-Verlag, 1979), pp. 53–75.

12 THEY WERE NUMERICAL RULES   A readable contemporary description by Lorenz of the problem of using equations to model the atmosphere is "Large-Scale Motions of the Atmosphere: Circulation," in *Advances in Earth Science*, ed. P. M. Hurley (Cambridge, Mass.: The M.I.T. Press, 1966), pp. 95–109. An early, influential analysis of this problem is L. F. Richardson, *Weather Prediction by Numerical Process* (Cambridge: Cambridge University Press, 1922).

13 PURITY OF MATHEMATICS   Lorenz. Also, an account of the conflicting pulls of mathematics and meteorology in his thinking is in "Irregularity: A Fundamental Property of the Atmosphere," Crafoord Prize Lecture presented at the Royal Swedish Academy of Sciences, Stockholm, Sept. 28, 1983, in *Tellus* 36A (1984), pp. 98–110.

14 "IT WOULD EMBRACE"   Pierre Simon de Laplace, *A Philosophical Essay on Probabilities* (New York: Dover, 1951).

15 "THE BASIC IDEA"   Winfree.

15 "THAT'S THE KIND OF RULE"   Lorenz.

16 SUDDENLY HE REALIZED   "On the Prevalence," p. 55.

17 SMALL ERRORS PROVED CATASTROPHIC   Of all the classical physicists and mathematicians who thought about dynamical systems, the one who best understood the possibility of chaos was Jules Henri Poincaré. Poincaré remarked in *Science and Method*:

"A very small cause which escapes our notice determines a considerable effect that we cannot fail to see, and then we say that the effect is due to chance. If we knew exactly the laws of nature and the situation of the universe at the initial moment, we could predict exactly the situation of that same universe at a succeeding moment. But even if it were the case that the natural laws had no longer any secret for us, we could still know the situation approximately. If that enabled us to predict the succeeding situation with the same approximation, that is all we require, and we should say that the phenomenon had been predicted, that it is governed by the laws. But it is not always so; it may happen that small differences in the initial conditions produce very great ones in the final phenomena. A small error in the former will produce an enormous error in the latter. Prediction becomes impossible. . . ."

Poincaré's warning at the turn of the century was virtually forgotten; in the United States, the only mathematician to seriously follow Poincaré's lead in the twenties and thirties was George D. Birkhoff, who, as it happened, briefly taught a young Edward Lorenz at M.I.T.

17 THAT FIRST DAY   Lorenz; also, "On the Prevalence," p. 56.

18 "WE CERTAINLY HADN'T"   Lorenz.

18 YEARS OF UNREAL OPTIMISM   Woods, Schneider; a broad survey of expert opinion at the time was "Weather Scientists Optimistic That New Findings Are Near," *The New York Times*, 9 September 1963, p. 1.

19 VON NEUMANN IMAGINED   Dyson.

19 VAST AND EXPENSIVE BUREAUCRACY   Bonner, Bengtsson, Woods, Leith.

20 FORECASTS OF ECONOMIC   Peter B. Medawar, "Expectation and Prediction," in *Pluto's Republic* (Oxford: Oxford University Press, 1982), pp. 301–4.

20 THE BUTTERFLY EFFECT   Lorenz originally used the image of a seagull; the more lasting name seems to have come from his paper, "Predictability: Does the Flap of a Butterfly's Wings in Brazil Set Off a Tornado in Texas?" address at the annual meeting of the American Association for the Advancement of Science in Washington, 29 December 1979.

21 SUPPOSE THE EARTH   Yorke.

21 "PREDICTION, NOTHING"   Lorenz, White.

22 THERE MUST BE A LINK   "The Mechanics of Vacillation."

23 FOR WANT OF A NAIL   George Herbert; cited in this context by Norbert Wiener, "Nonlinear Prediction and Dynamics," in *Collected Works with Commentaries*, ed. P. Masani (Cambridge, Mass.: The M.I.T. Press, 1981), 3:371. Wiener anticipated Lorenz in seeing at least the possibility of "self-amplitude of small details of the weather map." He noted, "A tornado is a highly local phenomenon, and apparent trifles of no great extent may determine its exact track."

24 "THE CHARACTER OF THE EQUATION"   John von Neumann, "Recent Theories of Turbulence" (1949), in *Collected Works*, ed. A. H. Taub (Oxford: Pergamon Press, 1963), 6:437.

24 CUP OF HOT COFFEE   "The predictability of hydrodynamic flow," in *Transactions of the New York Academy of Sciences* II:25:4 (1963), pp. 409–32.

25 "WE MIGHT HAVE TROUBLE"   Ibid., p. 410.

25 LORENZ TOOK A SET   This set of seven equations to model convection was devised by Barry Saltzman of Yale University, whom Lorenz was visiting. Usually the Saltzman equations behaved periodically, but one version "refused to settle down," as Lorenz said, and Lorenz realized that during this chaotic behavior four of the variables were approaching zero—thus they could be disregarded. Barry Saltzman, "Finite Amplitude Convection as an Initial Value Problem," *Journal of the Atmospheric Sciences* 19 (1962), p. 329.

29 GEODYNAMO   Malkus; the chaos view of the earth's magnetic fields is still hotly debated, with some scientists looking for other, ex-

ternal explanations, such as blows from huge meteorites. An early exposition of the idea that the reversals come from chaos built into the system is K. A. Robbins, "A moment equation description of magnetic reversals in the earth," *Proceedings of the National Academy of Science* 73 (1976), pp. 4297–4301.

29 WATER WHEEL   Malkus.
30 THREE EQUATIONS   This classic model, commonly called the Lorenz system, is:

$$dx/dt = 10(y - x)$$
$$dy/dt = -xz + 28x - y$$
$$dz/dt = xy - (8/3)z.$$

Since appearing in "Deterministic Nonperiodic Flow," the system has been widely analyzed; one authoritative technical volume is Colin Sparrow, *The Lorenz Equations, Bifurcations, Chaos, and Strange Attractors* (Springer-Verlag, 1982).

31 "ED, WE KNOW"   Malkus, Lorenz.
31 NO ONE THOUGHT   "Deterministic Nonperiod Flow" was cited about once a year in the mid 1960s by the scientific community; two decades later, it was cited more than one hundred times a year.

REVOLUTION

35 THE HISTORIAN OF SCIENCE   Kuhn's understanding of scientific revolutions has been widely dissected and debated in the twenty-five years since he put it forward, at about the time Lorenz was programming his computer to model weather. For Kuhn's views I have relied primarily on *The Structure of Scientific Revolutions*, 2nd ed. enl. (Chicago: University of Chicago Press, 1970) and secondarily on *The Essential Tension: Selected Studies in Scientific Tradition and Change* (Chicago: University of Chicago, 1977); "What Are Scientific Revolutions?" (Occasional Paper No. 18, Center for Cognitive Science, Massachusetts Institute of Technology); and Kuhn, interview. Another useful and important analysis of the subject is I. Bernard Cohen, *Revolution in Science* (Cambridge, Mass.: Belknap Press, 1985).

35 "I CAN'T MAKE   Structure, pp. 62–65, citing J. S. Bruner and Leo Postman, "On the Perception of Incongruity: A Paradigm," *Journal of Personality XVIII* (1949), p. 206.

36 MOPPING UP OPERATIONS   Structure, p. 24.
36 EXPERIMENTALISTS CARRY OUT   Tension, p. 229.
36 IN BENJAMIN FRANKLIN'S   Structure, pp. 13–15.
37 "UNDER NORMAL CONDITIONS   Tension, p. 234.
37 A PARTICLE PHYSICIST   Cvitanović
38 TOLSTOY   Ford, interview and "Chaos: Solving the Unsolvable, Predicting the Unpredictable," in *Chaotic Dynamics and Fractals*, ed. M. F. Barnsley and S. G. Demko (New York: Academic Press, 1985).

38 SUCH COINAGES   But Michael Berry notes that the OED has "Chaology (rare) 'the history or description of the chaos.' " Berry, "The

Unpredictable Bouncing Rotator: A Chaology Tutorial Machine,"
preprint, H. H. Wills Physics Laboratory, Bristol.

38 "IT's MASOCHISM Richter.

39 THESE RESULTS APPEAR J. Crutchfield, M. Nauenberg and J. Rud-
nick, "Scaling for External Noise at the Onset of Chaos," *Physical
Review Letters* 46 (1981), p. 933.

39 THE HEART OF CHAOS Alan Wolf, "Simplicity and Universality in
the Transition to Chaos," *Nature* 305 (1983), p. 182.

39 CHAOS NOW PRESAGES Joseph Ford, "What is Chaos, That We Should
Be Mindful of It?" preprint, Georgia Institute of Technology, Atlanta.

39 REVOLUTIONS DO NOT "What Are Scientific Revolutions?" p. 23.

39 "IT IS RATHER AS IF *Structure*, p. 111.

39 THE LABORATORY MOUSE Yorke and others.

40 WHEN ARISTOTLE LOOKED "What Are Scientific Revolutions?" pp.
2–10.

41 "IF TWO FRIENDS" Galileo *Opere* VIII: 277. Also VIII: 129–30.

42 "PHYSIOLOGICAL AND PSYCHIATRIC" David Tritton, "Chaos in the
swing of a pendulum," *New Scientist*, 24 July 1986, p. 37. This is
a readable, nontechnical essay on the philosophical implications
of pendulum chaos.

42 THAT CAN HAPPEN In practice, someone pushing a swing can always produce more or less regular motion, presumably using an
unconscious nonlinear feedback mechanism of his own.

42 YET, ODD AS IT SEEMS Among many analyses of the possible com-
plications of a simple driven pendulum, a good summary is D.
D'Humieres, M. R. Beasley, B. A. Huberman, and A. Libchaber,
"Chaotic States and Routes to Chaos in the Forced Pendulum,"
Physical Review A 26 (1982), pp. 3483–96.

43 SPACE BALLS Michael Berry researched the physics of this toy
both theoretically and experimentally. In "The Unpredictable
Bouncing Rotator" he describes a range of behaviors understandable only in the language of chaotic dynamics: "KAM tori, bifurcation of periodic orbits, Hamiltonian chaos, stable fixed points
and strange attractors."

44 FRENCH ASTRONOMER Hénon.

45 JAPANESE ELECTRICAL ENGINEER Ueda.

45 A YOUNG PHYSICIST Fox.

45 SMALE Smale, Yorke, Guckenheimer, Abraham, May, Feigenbaum; a brief, somewhat anecdotal account of Smale's thinking
during this period is "On How I Got Started in Dynamical Systems,"
in Steve Smale, *The Mathematics of Time: Essays on Dynamical
Systems, Economic Processes, and Related Topics* (New York:
Springer-Verlag, 1980), pp. 147–51.

45 THE SCENE IN MOSCOW Raymond H. Anderson, "Moscow Silences
a Critical American," *The New York Times*, 27 August 1966, p. 1;
Smale, "On the Steps of Moscow University," *The Mathematical
Intelligencer* 6:2, pp. 21–27.

46 WHEN HE RETURNED Smale.

48 A LETTER FROM A COLLEAGUE The colleague was N. Levinson.
Several threads of mathematics, running back to Poincaré, came

together here. The work of Birkhoff was one. In England, Mary Lucy Cartwright and J. E. Littlewood pursued the hints turned up by Balthasar van der Pol in chaotic oscillators. These mathematicians were all aware of the possibility of chaos in simple systems, but Smale, like most well-educated mathematicians, was unaware of their work, until the letter from Levinson.

48   ROBUST AND STRANGE   Smale; "On How I Got Started."
49   IT WAS JUST A VACUUM TUBE   van der Pol described his work in *Nature* 120 (1927), pp. 363–64.
49   "OFTEN AN IRREGULAR NOISE"   Ibid.
51   TO MAKE A SIMPLE   Smale's definitive mathematical exposition of this work is "Differentiable Dynamical Systems," *Bulletin of the American Mathematical Society* 1967, pp. 747–817 (also in *The Mathematics of Time*, pp. 1–82).
51   THE PROCESS MIMICS   Rössler.
52   BUT FOLDING   Yorke.
52   IT WAS A GOLDEN AGE   Guckenheimer, Abraham.
52   "IT'S THE PARADIGM SHIFT   Abraham.
53   A MODEST COSMIC MYSTERY   Marcus, Ingersoll, Williams; Philip S. Marcus, "Coherent Vortical Features in a Turbulent Two-Dimensional Flow and the Great Red Spot of Jupiter," paper presented at the 110th Meeting of the Acoustical Society of America, Nashville, Tennessee, 5 November 1985.
53   "THE RED SPOT ROARING"   John Updike, "The Moons of Jupiter," *Facing Nature* (New York: Knopf, 1985), p. 74.
54   VOYAGER HAD MADE   Ingersoll; also, Andrew P. Ingersoll, "Order from Chaos: The Atmospheres of Jupiter and Saturn," *Planetary Report* 4:3, pp. 8–11.
55   "YOU SEE THIS"   Marcus.
56   "GEE, WHAT ABOUT"   Marcus.

LIFE'S UPS AND DOWNS

59   RAVENOUS FISH   May, Schaffer, Yorke, Guckenheimer. May's famous review article on the lessons of chaos in population biology is "Simple Mathematical Models with Very Complicated Dynamics," *Nature* 261 (1976), pp. 459–67. Also: "Biological Populations with Nonoverlapping Generations: Stable Points, Stable Cycles, and Chaos," *Science* 186 (1974), pp. 645–47, and May and George F. Oster, "Bifurcations and Dynamic Complexity in Simple Ecological Models," *The American Naturalist* 110 (1976), pp. 573–99. An excellent survey of the development of mathematical modeling of populations, before chaos, is Sharon E. Kingsland, *Modeling Nature: Episodes in the History of Population Ecology* (Chicago: University of Chicago Press, 1985).
59   THE WORLD MAKES   May and Jon Seger, "Ideas in Ecology: Yesterday and Tomorrow," preprint, Princeton University, p. 25.
60   CARICATURES OF REALITY   May and George F. Oster, "Bifurcations and Dynamic Complexity in Simple Ecological Models," *The American Naturalist* 110 (1976), p. 573.

63 BY THE 1950S   May.
64 REFERENCE BOOKS   J. Maynard Smith, *Mathematical Ideas in Bi-ology* (Cambridge: Cambridge University Press, 1968), p. 18; Harvey J. Gold, *Mathematical Modeling of Biological Systems.*
65 IN THE BACK   May.
65 HE PRODUCED A REPORT   *Gonorrhea Transmission Dynamics and Control.* Herbert W. Hethcote and James A. Yorke (Berlin: Springer-Verlag, 1984).
65 THE EVEN-ODD SYSTEM   From computer simulations, Yorke found that the system forced drivers to make more trips to the filling station and to keep their tanks fuller all the time; thus the system increased the amount of gasoline sitting wastefully in the nation's automobiles at any moment.
65 HE ANALYZED THE MONUMENT'S SHADOW   Airport records later proved Yorke correct.
66 LORENZ'S PAPER   Yorke.
66 "FACULTY MEMBERS"   Murray Gell-Mann, "The Concept of the In-stitute," in *Emerging Syntheses in Science*, proceedings of the founding workshops of the Santa Fe Institute (Santa Fe: The Santa Fe Institute, 1985), p. 11.
67 HE GAVE A COPY   Yorke, Smale.
67 "IF YOU COULD WRITE"   Yorke.
68 HOW NONLINEAR NATURE IS   A readable essay on linearity, non-linearity, and the historical use of computers in understanding the difference is David Campbell, James P. Crutchfield, J. Doyne Farmer, and Erica Jen, "Experimental Mathematics: The Role of Compu-tation in Nonlinear Science," *Communications of the Association for Computing Machinery* 28 (1985), pp. 374–84.
68 "IT DOES NOT SAY"   Fermi, quoted in S. M. Ulam, *Adventures of a Mathematician* (New York: Scribners, 1976). Ulam also describes the origin of another important thread in the understanding of non-linearity, the Fermi-Pasta-Ulam theorem. Looking for problems that could be computed on the new MANIAC computer at Los Alamos, the scientists tried a dynamical system that was simply a vibrating string—a simple model "having, in addition, a physically correct small non-linear term." They found patterns coalescing into an unexpected periodicity. As Ulam recounts it: "The results were entirely different qualitatively from what even Fermi, with his great knowledge of wave motions, had expected. . . . To our surprise the string started playing a game of musical chairs, . . ." Fermi consid-ered the results unimportant, and they were not widely published, but a few mathematicians and physicists followed them up, and they became a particular part of the local lore at Los Alamos. *Ad-ventures*, pp. 226–28.
68 "NONELEPHANT ANIMALS"   quoted in "Experimental Mathemat-ics," p. 374.
68 "THE FIRST MESSAGE"   Yorke.
69 YORKE'S PAPER   Written with his student Tien-Yien Li. "Period Three Implies Chaos," *American Mathematical Monthly* 82 (1975), pp. 985–92.
69 MAY CAME TO BIOLOGY   May.

69   "WHAT THE CHRIST"   May; it was this seemingly unanswerable question that drove him from analytic methods to numerical experimentation, meant to provide intuition, at least.
73   STARTLING THOUGH IT WAS   Yorke.
74   A. N. SARKOVSKII   "Coexistence of Cycles of a Continuous Map of a Line into Itself," *Ukrainian Mathematics Journal* 16 (1964), p. 61.
76   SOVIET MATHEMATICIANS AND PHYSICISTS   Sinai, personal communication, 8 December 1986.
76   SOME WESTERN CHAOS EXPERTS   e.g., Feigenbaum, Cvitanović.
77   TO SEE DEEPER   Hoppensteadt, May.
77   THE FEELING OF ASTONISHMENT   Hoppensteadt.
78   WITHIN ECOLOGY   May.
79   NEW YORK CITY MEASLES   William M. Schaffer and Mark Kot, "Nearly One-dimensional Dynamics in an Epidemic," *Journal of Theoretical Biology* 112 (1985), pp. 403–27; Schaffer, "Stretching and Folding in Lynx Fur Returns: Evidence for a Strange Attractor in Nature," *The American Naturalist* 124 (1984), pp. 798–820.
80   THE WORLD WOULD BE   "Simple Mathematical Models," p. 467.
80   "THE MATHEMATICAL INTUITION"   Ibid.

A GEOMETRY OF NATURE

83   A PICTURE OF REALITY   Mandelbrot, Gomory, Voss, Barnsley, Richter, Mumford, Hubbard, Shlesinger. The Benoit Mandelbrot bible is *The Fractal Geometry of Nature* (New York: Freeman, 1977). An interview by Anthony Barcellos appears in *Mathematical People,* ed. Donald J. Albers and G. L. Alexanderson (Boston: Birkhäuser, 1985). Two essays by Mandelbrot that are less well known and extremely interesting are "On Fractal Geometry and a Few of the Mathematical Questions It Has Raised," *Proceedings of the International Congress of Mathematicians,* 16–14 August 1983, Warsaw, pp. 1661–75; and "Towards a Second Stage of Indeterminism in Science," preprint, IBM Thomas J. Watson Research Center, Yorktown Heights, New York. Review articles on applications of fractals have grown too common to list, but two useful examples are Leonard M. Sander, "Fractal Growth Processes," *Nature* 322 (1986), pp. 789–93; Richard Voss, "Random Fractal Forgeries: From Mountains to Music," in *Science and Uncertainty,* ed. Sara Nash (London: IBM United Kingdom, 1985).
83   CHARTED ON THE OLDER MAN'S BLACKBOARD   Houthakker, Mandelbrot.
84   WASSILY LEONTIEF   Quoted in *Fractal Geometry,* p. 423.
86   INTRODUCED FOR A LECTURE   Woods Hole Oceanographic Institute, August 1985.
87   BORN IN WARSAW   Mandelbrot.
88   BOURBAKI   Mandelbrot, Richter. Little has been written about Bourbaki even now; one playful introduction is Paul R. Halmos, "Nicholas Bourbaki," *Scientific American* 196 (1957), pp. 88–89.
89   MATHEMATICS SHOULD BE SOMETHING   Smale.

89   THE FIELD DEVELOPS   Peitgen.
90   PIONEER-BY-NECESSITY   "Second Stage," p. 5.
92   THIS HIGHLY ABSTRACT   Mandelbrot; *Fractal Geometry*, p. 74; J. M. Berger and Benoit Mandelbrot, "A New Model for the Clustering of Errors on Telephone Circuits," *IBM Journal of Research and Development* 7 (1963), pp. 224–36.
93   THE JOSEPH EFFECT   *Fractal Geometry*, p. 248.
94   CLOUDS ARE NOT SPHERES   Ibid., p. 1, for example.
95   WONDERING ABOUT COASTLINES   Ibid., p. 27.
97   THE PROCESS OF ABSTRACTION   Ibid., p. 17.
97   "THE NOTION"   Ibid., p. 18.
98   ONE WINTRY AFTERNOON   Mandelbrot.
100 THE EIFFEL TOWER   *Fractal Geometry*, p. 131, and "On Fractal Geometry," p. 1663.
102 ORIGINATED BY MATHEMATICIANS   F. Hausdorff and A. S. Besicovich.
102 "THERE WAS A LONG HIATUS"   Mandelbrot.
103 IN THE NORTHEASTERN   Scholz; C. H. Scholz and C. A. Aviles, "The Fractal Geometry of Faults and Faulting," preprint, Lamont-Doherty Geophysical Observatory; C. H. Scholz, "Scaling Laws for Large Earthquakes," *Bulletin of the Seismological Society of America* 72 (1982), pp. 1–14.
104 "A MANIFESTO"   *Fractal Geometry*, p. 24.
104 "NOT A HOW-TO BOOK"   Scholz.
107 "IT'S A SINGLE MODEL"   Scholz.
109 "IN THE GRADUAL"   William Bloom and Don W. Fawcett, *A Textbook of Histology* (Philadelphia: W. B. Saunders, 1975).
109 SOME THEORETICAL BIOLOGISTS   One review of these ideas is Ary L. Goldberger, "Nonlinear Dynamics, Fractals, Cardiac Physiology, and Sudden Death," in *Temporal Disorder in Human Oscillatory Systems*, ed. L. Rensing, U. An der Heiden, M. Mackey (New York: Springer-Verlag, 1987).
109 THE NETWORK OF SPECIAL FIBERS   Goldberger, West.
109 SEVERAL CHAOS-MINDED CARDIOLOGISTS   Ary L. Goldberger, Valmik Bhargava, Bruce J. West and Arnold J. Mandell, "On a Mechanism of Cardiac Electrical Stability: The Fractal Hypothesis," *Biophysics Journal* 48 (1985), p. 525.
110 WHEN E. I. DUPONT   Barnaby J. Feder, "The Army May Have Matched the Goose," *The New York Times*, 30 November 1986, 4:16.
110 "I STARTED LOOKING"   Mandelbrot.
111 HIS NAME APPEARED   I. Bernard Cohen, *Revolution in Science* (Cambridge, Mass.: Belknap, 1985), p. 46.
111 "OF COURSE, HE IS A BIT"   Mumford.
111 "HE HAD SO MANY DIFFICULTIES"   Richter.
112 IF THEY WANTED TO AVOID   Just as Mandelbrot later could avoid the credit routinely given to Mitchell Feigenbaum in references to *Feigenbaum numbers* and *Feigenbaum universality*. Instead, Mandelbrot habitually referred to P. J. Myrberg, a mathematician who had studied iterates of quadratic mappings in the early 1960s, obscurely.

112   "Mandelbrot didn't have everybody's"   Richter.
113   "The politics affected"   Mandelbrot.
114   Exxon's huge research facility   Klafter.
114   One mathematician told friends   Related by Huberman.
117   "Why is it that"   "Freedom, Science, and Aesthetics," in *Schönheit im Chaos*, p. 35.
117   "The period had no sympathy"   John Fowles, *A Maggot* (Boston: Little, Brown, 1985), p. 11.
118   "We have the astronomers"   Robert H. G. Helleman, "Self-Generated Behavior in Nonlinear Mechanics," in *Fundamental Problems in Statistical Mechanics* 5, ed. E. G. D. Cohen (Amsterdam: North-Holland, 1980), p. 165.
118   But physicists wanted more   Leo Kadanoff, for example, asked "Where is the physics of fractals?" in *Physics Today*, February 1986, p. 6, and then answered the question with a new "multifractal" approach in *Physics Today*, April 1986, p. 17, provoking a typically annoyed response from Mandelbrot, *Physics Today*, September 1986, p. 11. Kadanoff's theory, Mandelbrot wrote, "fills me with the pride of a father—soon to be a grandfather?"

STRANGE ATTRACTORS

121   The great physicists   Ruelle, Hénon, Rössler, Sinai, Feigenbaum, Mandelbrot, Ford, Kraichnan. Many perspectives exist on the historical context for the strange-attractor view of turbulence. A worthwhile introduction is John Miles, "Strange Attractors in Fluid Dynamics," in *Advances in Applied Mechanics* 24 (1984), pp. 189–214. Ruelle's most accessible review article is "Strange Attractors," *Mathematical Intelligencer* 2 (1980), pp. 126–37; his catalyzing proposal was David Ruelle and Floris Takens, "On the Nature of Turbulence," *Communications in Mathematical Physics* 20 (1971), pp. 167–92; his other essential papers include "Turbulent Dynamical Systems," *Proceedings of the International Congress of Mathematicians, 16–24 August 1983, Warsaw*, pp. 271–86; "Five Turbulent Problems," *Physica* 7D (1983), pp. 40–44; and "The Lorenz Attractor and the Problem of Turbulence," in *Lecture Notes in Mathematics No. 565* (Berlin: Springer-Verlag, 1976), pp. 146–58.
121   There was a story   Many versions of this exist. Orszag cites four substitutes for Heisenberg—von Neumann, Lamb, Sommerfeld, and von Karman—and adds, "I imagine if God actually gave an answer to these four people it would be different in each case."
123   This assumption   Ruelle; also "Turbulent Dynamical Systems," p. 281.
123   text on fluid dynamics   L. D. Landau and E. M. Lifshitz, *Fluid Mechanics* (Oxford: Pergamon, 1959).
124   the oscillatory, the skewed varicose   Malkus.
126   "That's true"   Swinney.
128   In 1973 Swinney   Swinney, Gollub.
130   "It was a string-and-sealing-wax"   Dyson.

131 "So we read that" Swinney.
131 When they began reporting Swinney, Gollub.
131 "There was the transition" Swinney.
131 Experiment failed to confirm J. P. Gollub and H. L. Swinney, "Onset of Turbulence in a Rotating Fluid," *Physical Review Letters* 35 (1975), p. 927. These first experiments only opened the door to an appreciation of the complex spatial behaviors that could be produced by varying the few parameters of flow between rotating cylinders. The next few years identified patterns from "corkscrew wavelets" to "wavy inflow and outflow" to "interpenetrating spirals." A summary is C. David Andereck, S. S. Liu, and Harry L. Swinney, "Flow Regimes in a Circular Couette System with Independently Rotating Cylinders," *Journal of Fluid Mechanics* 164 (1986), pp. 155–83.
132 David Ruelle sometimes said Ruelle.
132 "Always nonspecialists find" Ruelle.
133 He wrote a paper "On the Nature of Turbulence."
133 Opinions still varied They quickly discovered that some of their ideas had already appeared in the Russian literature; "on the other hand, the mathematical interpretation which we give of turbulence seems to remain our own responsibility!" they wrote. "Note Concerning Our Paper 'On the Nature of Turbulence,' " *Communications in Mathematical Physics* 23 (1971), pp. 343–44.
133 psychoanalytically "suggestive" Ruelle.
133 "Did you ever ask God" "Strange Attractors," p. 131.
133 "Takens happened" Ruelle.
135 "Some mathematicians in California" Ralph H. Abraham and Christopher D. Shaw, *Dynamics: The Geometry of Behavior* (Santa Cruz: Aerial: 1984).
137 "It always bothers me" Richard P. Feynman, *The Character of Physical Law* (Cambridge, Mass.: The M.I.T. Press, 1967), p. 57.
138 David Ruelle suspected Ruelle.
139 The reaction of the scientific public "Turbulent Dynamical Systems," p. 275.
139 Edward Lorenz had attached "Deterministic Nonperiodic Flow," p. 137.
140 "It is difficult to reconcile Ibid., p. 140.
141 He went to visit Lorenz Ruelle.
141 "Don't form a selfish concept Ueda reviews his early discoveries from the point of view of electrical circuits in "Random Phenomena Resulting from Nonlinearity in the System Described by Duffing's Equation," in *International Journal of Non-Linear Mechanics* 20 (1985), pp. 481–91, and gives a personal account of his motivation and the cool response of his colleagues in a postscript. Also, Stewart, private communication.
141 "a sausage in a sausage" Rössler.
144 The most illuminating strange attractor Hénon; he reported his invention in "A Two-Dimensional Mapping with a Strange Attractor," in *Communications in Mathematical Physics* 50 (1976), pp. 69–77, and Michel Hénon and Yves Pomeau, "Two Strange Attractors with a Simple Structure," in *Turbulence and the Navier-*

Stokes Equations, ed. R. Teman (New York: Springer-Verlag, 1977).
145   IS THE SOLAR SYSTEM   Wisdom.
146   "TO HAVE MORE FREEDOM"   Michel Hénon and Carl Heiles, "The Applicability of the Third Integral of Motion: Some Numerical Experiments," *Astronomical Journal* 69 (1964), p. 73.
147   AT THE OBSERVATORY   Hénon.
147   "I, TOO, was convinced"   Hénon.
147   "HERE COMES THE SURPRISE"   "The Applicability," p. 76.
149   "BUT THE MATHEMATICAL APPROACH"   Ibid., p. 79.
149   A VISITING PHYSICIST   Yves Pomeau.
149   "SOMETIMES ASTRONOMERS ARE FEARFUL"   Hénon.
152   OTHERS ASSEMBLED MILLIONS   Ramsey.
153   "I HAVE NOT SPOKEN"   "Strange Attractors," p. 137.

UNIVERSALITY

157   "YOU CAN FOCUS"   Feigenbaum. Feigenbaum's crucial papers on universality are "Quantitative Unversality for a Class of Nonlinear Transformations," *Journal of Statistical Physics* 19 (1978), pp. 25–52, and "The Universal Metric Properties of Nonlinear Transformations," *Journal of Statistical Physics* 21 (1979), pp. 669–706; a somewhat more accessible presentation, though still requiring some mathematics, is his review article, "Universal Behavior in Nonlinear Systems," *Los Alamos Science* 1 (Summer 1981), pp. 4–27. I also relied on his unpublished recollections, "The Discovery of Universality in Period Doubling."
157   WHEN FEIGENBAUM CAME TO LOS ALAMOS   Feigenbaum, Carruthers, Cvitanović, Campbell, Farmer, Visscher, Kerr, Hasslacher, Jen.
158   "IF YOU HAD SET UP"   Carruthers.
159   THE MYSTERY OF THE UNIVERSE   Feigenbaum.
160   OCCASIONALLY AN ADVISOR   Carruthers.
160   AS KADANOFF VIEWED   Kadanoff.
163   "THE CEASELESS MOTION"   Gustav Mahler, letter to Max Marschalk.
164   "WITH LIGHT POISE"   Goethe's *Zür Farbenlehre* is now available in several editions. I relied on the beautifully illustrated *Goethe's Color Theory*, ed. Rupprecht Matthaei, trans. Herb Aach (New York: Van Nostrand Reinhold, 1970); more readily available is *Theory of Colors* (Cambridge, Mass.: The M.I.T. Press, 1970), with an excellent introduction by Deane B. Judd.
167   THIS ONE INNOCENT-LOOKING EQUATION   At one point, Ulam and von Neumann used its chaotic properties as a solution to the problem of generating random numbers with a finite digital computer.
167   TO METROPOLIS, STEIN, AND STEIN   This paper—the sole pathway from Stanislaw Ulam and John von Neumann to James Yorke and Mitchell Feigenbaum—is "On Finite Limit Sets for Transformations on the Unit Interval," *Journal of Combinatorial Theory* 15 (1973), pp. 25–44.
168   DOES A CLIMATE EXIST   "The Problem of Deducing the Climate from the Governing Equations," *Tellus* 16 (1964), pp. 1–11.

170 THE WHITE EARTH CLIMATE   Manabe.
171 HE KNEW NOTHING OF LORENZ   Feigenbaum.
172 ODDLY   May.
173 THE SAME COMBINATIONS OF R'S AND L'S   "On Finite Limit Sets," pp. 30–31. The crucial hint: "The fact that these patterns . . . are a common property of four apparently unrelated transformations . . . suggests that the pattern sequence is a general property of a wide class of mappings. For this reason we have called this sequence of patterns the *U*-sequence where '*U*' stands (with some exaggeration) for 'universal.' " But the mathematicians never imagined that the universality would extend to actual numbers; they made a table of 84 different parameter values, each taken to seven decimal places, without observing the geometrical relationships hidden there.
174 "THE WHOLE TRADITION OF PHYSICS"   Feigenbaum.
179 HIS FRIENDS SPECULATED   Cvitanović.
180 SUDDENLY YOU COULD SEE   Ford.
180 PRIZES AND AWARDS   The MacArthur fellowship; the 1986 Wolf Prize in physics.
182 "FEIGENBAUMOLOGY"   Dyson.
183 "IT WAS A VERY HAPPY"   Gilmore.
183 BUT ALL THE WHILE   Cvitanović.
183 WORK BY OSCAR E. LANFORD   Even then, the proof was unorthodox in that it depended on tremendous amounts of numerical calculation, so that it could not be carried out or checked without the use of a computer. Lanford; Oscar E. Lanford, "A Computer-Assisted Proof of the Feigenbaum Conjectures," *Bulletin of the American Mathematical Society* 6 (1982), p. 427; also, P. Collet, J.-P. Eckmann, and O. E. Lanford, "Universal Properties of Maps on an Interval," *Communications in Mathematical Physics* 81 (1980), p. 211.
184 "SIR, DO YOU MEAN"   Feigenbaum; "The Discovery of Universality," p. 17.
184 IN THE SUMMER OF 1977   Ford, Feigenbaum, Lebowitz.
184 "MITCH HAD SEEN UNIVERSALITY"   Ford.
184 "SOMETHING DRAMATIC HAPPENED"   Feigenbaum.

THE EXPERIMENTER

191 "ALBERT IS GETTING MATURE"   Libchaber, Kadanoff.
191 HE SURVIVED THE WAR   Libchaber.
192 "HELIUM IN A SMALL BOX"   Albert Libchaber, "Experimental Study of Hydrodynamic Instabilities. Rayleigh-Benard Experiment: Helium in a Small Box," in *Nonlinear Phenomena at Phase Transitions and Instabilities*, ed. T. Riste (New York: Plenum, 1982), p. 259.
192 THE LABORATORY OCCUPIED   Libchaber, Feigenbaum.
195 "SCIENCE WAS CONSTRUCTED"   Libchaber.
195 "BUT YOU KNOW THEY DO!"   Libchaber.
196 "THE FLECKED RIVER"   Wallace Stevens, "This Solitude of Cata-

racts," *The Palm at the End of the Mind*, ed. Holly Stevens (New York: Vintage, 1972), p. 321.

196 "INSOLID BILLOWING OF THE SOLID"   "Reality Is an Activity of the Most August Imagination," Ibid., p. 396.

197 "BUILDS ITS OWN BANKS"   Theodor Schwenk, *Sensitive Chaos* (New York: Schocken, 1976), p. 19.

198 "ARCHETYPAL PRINCIPLE"   Ibid.

198 "THIS PICTURE OF STRANDS"   Ibid., p. 16.

198 "THE INEQUALITIES"   Ibid., p. 39.

198 "IT MAY BE"   D'Arcy Wentworth Thompson, *On Growth and Form*, J. T. Bonner, ed. (Cambridge: Cambridge University Press, 1961), p. 8.

200 "BEYOND COMPARISON THE FINEST"   Ibid., p. viii.

200 "FEW HAD ASKED"   Stephen Jay Gould, *Hen's Teeth and Horse's Toes* (New York: Norton, 1983), p. 369.

202 "DEEP-SEATED RHYTHMS OF GROWTH"   *On Growth and Form*, p. 267.

202 "THE INTERPRETATION IN TERMS OF FORCE"   Ibid., p. 114.

204 IT WAS SO SENSITIVE   Campbell.

204 "IT WAS CLASSICAL PHYSICS"   Libchaber.

205 NOW, HOWEVER, A NEW FREQUENCY   Libchaber and Maurer, 1980 and 1981. Also Cvitanović's introduction gives a lucid summary.

208 "THE NOTION THAT THE ACTUAL"   Hohenberg.

208 "THEY STOOD AMID THE SCATTERED"   Feigenbaum, Libchaber.

209 "YOU HAVE TO REGARD IT"   Gollub.

209 A VAST BESTIARY OF LABORATORY EXPERIMENTS   The literature is equally vast. One summary of the early melding of theory and experiment in a variety of systems is Harry L. Swinney, "Observations of Order and Chaos in Nonlinear Systems," *Physica* 7D (1983), pp. 3–15; Swinney provides a list of references divided into categories, from electronic and chemical oscillators to more esoteric kinds of experiments.

209 TO MANY, EVEN MORE CONVINCING   Valter Franceschini and Claudio Tebaldi, "Sequences of Infinite Bifurcations and Turbulence in a Five-Mode Truncation of the Navier-Stokes Equations," *Journal of Statistical Physics* 21 (1979), pp. 707–26.

209 IN 1980 A EUROPEAN GROUP   P. Collet, J.-P. Eckmann, and H. Koch, "Period Doubling Bifurcations for Families of Maps on R^n," *Journal of Statistical Physics* 25 (1981), p. 1.

210 "A PHYSICIST WOULD ASK ME"   Libchaber.

IMAGES OF CHAOS

215 MICHAEL BARNSLEY MET   Barnsley.

216 RUELLE SHUNTED IT BACK   Barnsley.

217 JOHN HUBBARD, AN AMERICAN   Hubbard; also Adrien Douady, "Julia Sets and the Mandelbrot Set," in pp. 161–73. The main text of *The Beauty of Fractals* also give a mathematical summary of Newton's method, as well as the other meeting grounds of complex dynamics discussed in this chapter.

217 "NOW, FOR EQUATIONS" "Julia Sets and the Mandelbrot Set," p. 170.

217 HE STILL PRESUMED Hubbard.

219 A BOUNDARY BETWEEN TWO COLORS Hubbard; *The Beauty of Fractals;* Peter H. Richter and Heinz-Otto Peitgen, "Morphology of Complex Boundaries," *Bunsen-Gesellschaft für Physikalische Chemie* 89 1985), pp. 575–88.

221 THE MANDELBROT SET A readable introduction, with instructions for writing a do-it-yourself microcomputer program, is A. K. Dewdney, "Computer Recreations," *Scientific American* (August 1985), pp. 16–32. Peitgen and Richter in *The Beauty of Fractals* offer a detailed review of the mathematics, as well as some of the most spectacular pictures available.

221 THE MOST COMPLEX OBJECT Hubbard, for example.

221 "YOU OBTAIN AN INCREDIBLE VARIETY "Julia Sets and the Mandelbrot Set," p. 161.

222 IN 1979 MANDELBROT DISCOVERED Mandelbrot, Laff, Hubbard. A first-person account by Mandelbrot is "Fractals and the Rebirth of Iteration Theory," in *The Beauty of Fractals*, pp. 151–60.

223 AS HE TRIED CALCULATING Mandelbrot; *The Beauty of Fractals.*

228 MANDELBROT STARTED WORRYING Mandelbrot.

228 NO TWO PIECES ARE "TOGETHER" Hubbard.

229 "EVERYTHING WAS VERY GEOMETRIC" Peitgen.

229 AT CORNELL, MEANWHILE Hubbard.

229 RICHTER HAD COME TO COMPLEX SYSTEMS Richter.

230 "IN A BRAND NEW AREA" Peitgen.

231 "RIGOR IS THE STRENGTH" Peitgen.

233 FRACTAL BASIN BOUNDARIES Yorke; a good introduction, for the technically inclined, is Steven W. MacDonald, Celso Grebogi, Edward Ott, and James A. Yorke, "Fractal Basin Boundaries," *Physica* 17D (1985), pp. 125–83.

233 AN IMAGINARY PINBALL MACHINE Yorke.

234 "NOBODY CAN SAY" Yorke, remarks at Conference on Perspectives in Biological Dynamics and Theoretical Medicine, National Institutes of Health, Bethesda, Maryland, 10 April 1986.

235 TYPICALLY, MORE THAN THREE-QUARTERS Yorke.

235 THE BORDER BETWEEN CALM AND CATASTROPHE Similarly, in a text meant to introduce chaos to engineers, H. Bruce Stewart and J. M. Thompson warned: "Lulled into a false sense of security by his familiarity with the *unique* response of a linear system, the busy analyst or experimentalist shouts 'Eureka, this is the solution,' once a simulation settles onto an equilibrium of steady cycle, without bothering to explore patiently the outcome from different starting conditions. To avoid potentially dangerous errors and disasters, industrial designers must be prepared to devote a greater percentage of their effort into exploring the full range of dynamic responses of their systems." *Nonlinear Dynamics and Chaos* (Chichester: Wiley, 1986), p. xiii.

236 "PERHAPS WE SHOULD BELIEVE" *The Beauty of Fractals*, p. 136.

236 WHEN HE WROTE ABOUT e.g., "Iterated Function Systems and the

Global Construction of Fractals," *Proceedings of the Royal Society of London* A 399 (1985), pp. 243–75.

238 "IF THE IMAGE IS COMPLICATED" Barnsley.
239 "THERE IS NO RANDOMNESS" Hubbard.
239 "RANDOMNESS IS A RED" Barnsley.

THE DYNAMICAL SYSTEMS COLLECTIVE

243 SANTA CRUZ Farmer, Shaw, Crutchfield, Packard, Burke, Nauenberg, Abrahams, Guckenheimer. The essential Robert Shaw, applying information theory to chaos, is *The Dripping Faucet as a Model Chaotic System* (Santa Cruz: Aerial, 1984), along with "Strange Attractors, Chaotic Behavior, and Information Theory," *Zeitschrift für Naturforschung* 36a (1981), p. 80. An account of the roulette adventures of some of the Santa Cruz students, conveying much of the color of these years, is Thomas Bass, *The Eudemonic Pie* (Boston: Houghton Mifflin, 1985).

244 HE DID NOT KNOW Shaw.
244 WILLIAM BURKE, a SANTA CRUZ COSMOLOGIST Burke, Spiegel.
245 "COSMIC ARRHYTHMIAS" Edward A. Spiegel, "Cosmic Arrhythmias," in *Chaos in Astrophysics*, J. R. Buchler et al., eds. (New York: D. Reidel, 1985), pp. 91–135.
245 THE ORIGINAL PLANS Farmer, Crutchfield.
246 BY BUILDING UP Shaw, Crutchfield, Burke.
246 A FEW MINUTES LATER Shaw.
247 "ALL YOU HAVE TO DO" Abraham.
248 DOYNE FARMER Farmer is the main figure and Packard is a secondary figure in *The Eudemonic Pie*, the story of the roulette project, written by a sometime associate of the group.
249 PHYSICS AT SANTA CRUZ Burke, Farmer, Crutchfield.
250 "GIZMO-ORIENTED" Shaw.
252 FORD HAD ALREADY DECIDED Ford.
252 THEY REALIZED THAT MANY SORTS Shaw, Farmer.
255 INFORMATION THEORY The classic text, still quite readable, is Claude E. Shannon and Warren Weaver, *The Mathematical Theory of Communication* (Urbana: University of Illinois, 1963), with a helpful introduction by Weaver.
257 "WHEN ONE MEETS THE CONCEPT" Ibid., p. 13.
258 NORMAN PACKARD WAS READING Packard.
258 IN DECEMBER 1977 Shaw.
259 WHEN LORENZ WALKED INTO THE ROOM Shaw, Farmer.
259 HE FINALLY MAILED HIS PAPER "Strange Attractors, Chaotic Behavior, and Information Flow."
261 A. N. KOLMOGOROV AND YASHA SINAI Sinai, private communication.
261 AT THE PINNACLE Packard.
262 "YOU DON'T SEE SOMETHING" Shaw.
262 "IT'S A SIMPLE EXAMPLE" Shaw.
263 SYSTEMS THAT THE SANTA CRUZ GROUP Farmer; a dynamical sys-

tems approach to the immune system, modeling the human body's ability to "remember" and to recognize patterns creatively, is outlined in J. Doyne Farmer, Norman H. Packard, and Alan S. Perelson, "The Immune System, Adaptation, and Machine Learning," preprint, Los Alamos National Laboratory, 1986.

263 ONE IMPORTANT VARIABLE   *The Dripping Faucet*, p. 4.

263 "A STATE-OF-THE-ART COMPUTER CALCULATION"   Ibid.

264 A "PSEUDOCOLLOQUIUM"   Crutchfield.

265 "IT TURNS OUT"   Shaw.

266 "WHEN YOU THINK ABOUT A VARIABLE"   Farmer.

266 RECONSTRUCTING THE PHASE SPACE   These methods, which became a mainstay of experimental technique in many different fields, were greatly refined and extended by the Santa Cruz researchers and other experimentalists and theorists. One of the key Santa Cruz proposals was Norman H. Packard, James P. Crutchfield, J. Doyne Farmer, and Robert S. Shaw [the canonical byline list], "Geometry from a Time Series," *Physical Review Letters* 47 (1980), p. 712. The most influential paper on the subject by Floris Takens was "Detecting Strange Attractors in Turbulence," in *Lecture Notes in Mathematics* 898, D. A. Rand and L. S. Young, eds. (Berlin: Springer-Verlag, 1981), p. 336. An early but fairly broad review of the techniques of reconstructing phase-space portraits is Harold Froehling, James P. Crutchfield, J. Doyne Farmer, Norman H. Packard, and Robert S. Shaw, "On Determining the Dimension of Chaotic Flows," *Physica* 3D (1981), pp. 605–17.

267 "GOD, WE'RE STILL"   Crutchfield.

267 SOME PROFESSORS DENIED   e.g., Nauenberg.

267 "WE HAD NO ADVISOR"   Shaw.

268 MORE INTERESTED IN REAL SYSTEMS   Not that the students ignored maps altogether. Crutchfield, inspired by May's work, spent so much time in 1978 making bifurcation diagrams that he was barred from the computer center's plotter. Too many pens had been destroyed laying down the thousands of dots.

268 LANFORD LISTENED POLITELY   Farmer.

268 "IT WAS MY NAIVETÉ"   Farmer.

269 "AUDIOVISUAL AIDS"   Shaw.

270 ONE DAY BERNARDO HUBERMAN   Crutchfield, Huberman.

270 "IT WAS ALL VERY VAGUE"   Huberman.

270 THE FIRST PAPER   Bernardo A. Huberman and James P. Crutchfield, "Chaotic States of Anharmonic Systems in Periodic Fields," *Physical Review Letters* 43 (1979), p. 1743.

270 FARMER WAS ANGERED   Crutchfield.

271 CLIMATE SPECIALISTS   This is a continuing debate in the journal *Nature*, for example.

271 ECONOMISTS ANALYZING STOCK MARKET   Ramsey.

271 FRACTAL DIMENSION, HAUSDORFF DIMENSION   J. Doyne Farmer, Edward Ott, and James A. Yorke, "The Dimension of Chaotic Attractors," *Physica* 7D (1983), pp. 153–80.

271 "THE FIRST LEVEL OF KNOWLEDGE"   Ibid., p. 154.

INNER RHYTHMS

275 HUBERMAN LOOKED OUT   Huberman, Mandell (interviews and remarks at Conference on Perspectives in Biological Dynamics and Theoretical Medicine, Bethesda, Maryland, 11 April 1986). Also, Bernardo A. Huberman, "A Model for Dysfunctions in Smooth Pursuit Eye Movement," preprint, Xerox Palo Alto Research Center, Palo Alto, California.

279 "THREE THINGS HAPPEN"   Abraham. The basic introduction to the Gaia hypothesis—an imaginative dynamical view of how the earth's complex systems regulate themselves, somewhat sabotaged by its deliberate anthropomorphism—is J. E. Lovelock, *Gaia: A New Look at Life on Earth* (Oxford: Oxford University Press, 1979).

280 RESEARCHERS INCREASINGLY RECOGNIZED   A somewhat arbitrary selection of references on physiological topics (each with useful citations of its own): Ary L. Goldberger, Valmik Bhargava, and Bruce J. West, "Nonlinear Dynamics of the Heartbeat," *Physica* 17D (1985), pp. 207–14. Michael C. Mackay and Leon Glass, "Oscillation and Chaos in Physiological Control Systems," *Science* 197 (1977), p. 287. Mitchell Lewis and D. C. Rees, "Fractal Surfaces of Proteins," *Science* 230 (1985), pp. 1163–65. Ary L. Goldberger, et al., "Nonlinear Dynamics in Heart Failure: Implications of Long-Wavelength Cardiopulmonary Oscillations," *American Heart Journal* 107 (1984), pp. 612–15. Teresa Ree Chay and John Rinzel, "Bursting, Beating, and Chaos in an Excitable Membrane Model," *Biophysical Journal* 47 (1985), pp. 357–66. A particularly useful and wide-ranging collection of other such papers is *Chaos*, Arun V. Holden, ed. (Manchester: Manchester University Press, 1986).

280 "A DYNAMICAL SYSTEM OF VITAL INTEREST"   Ruelle, "Strange Attractors," p. 48.

281 "IT'S TREATED BY PHYSICIANS"   Glass.

281 "WE'RE AT A NEW FRONTIER"   Goldberger.

283 MATHEMATICIANS AT THE COURANT INSTITUTE   Peskin; David M. McQueen and Charles S. Peskin, "Computer-Assisted Design of Pivoting Disc Prosthetic Mitral Valves," *Journal of Thoracic and Cardiovascular Surgery* 86 (1983), pp. 126–35.

283 A PATIENT WITH A SEEMINGLY HEALTHY HEART   Cohen.

284 "THE BUSINESS OF DETERMINING"   Winfree.

285 A STRONG SENSE OF GEOMETRY   Winfree develops his view of geometric time in biological systems in a provocative and beautiful book, *When Time Breaks Down: The Three-Dimensional Dynamics of Electrochemical Waves and Cardiac Arrhythmias* (Princeton: Princeton University Press, 1987); a review article on the applications to heart rhythms is Arthur T. Winfree, "Sudden Cardiac Death: A Problem in Topology," *Scientific American* 248 (May 1983), p. 144.

285 "I HAD A HEADFUL"   Winfree.

286 "YOU GO TO A MOSQUITO"   Winfree.

288 SHE REPORTED FEELING GREAT   Strogatz; Charles A. Czeisler, et al., "Bright Light Resets the Human Circadian Pacemaker Independent of the Timing of the Sleep-Wake Cycle," *Science* 233 (1986), pp.

667–70. Steven Strogatz, "A Comparative Analysis of Models of the Human Sleep-Wake Cycle," preprint, Harvard University, Cambridge, Massachusetts.

288 HE HAD GAINED Winfree.
288 "WHEN MINES DECIDED" "Sudden Cardiac Death."
288 TO DO SO, HOWEVER Ideker.
289 "THE CARDIAC EQUIVALENT" Winfree.
289 IDEKER'S IMMEDIATE INTENTION Ideker.
290 THEY USED TINY AGGREGATES Glass.
290 "EXOTIC DYNAMIC BEHAVIOR" Michael R. Guevara, Leon Glass, and Alvin Schrier, "Phase Locking, Period-Doubling Bifurcations, and Irregular Dynamics in Periodically Stimulated Cardiac Cells," *Science* 214 (1981), p. 1350.
290 "MANY DIFFERENT RHYTHMS" Glass.
290 "IT IS A CLEAR INSTANCE" Cohen.
291 "PEOPLE HAVE MADE THESE WEIRD" Glass.
292 "DYNAMICAL THINGS ARE GENERALLY" Winfree.
292 "SYSTEMS THAT NORMALLY OSCILLATE" Leon Glass and Michael C. Mackay, "Pathological Conditions Resulting from Instabilities in Physiological Control Systems," *Annals of the New York Academy of Sciences* 316 (1979), p. 214.
293 "FRACTAL PROCESSES" Ary L. Goldberger, Valmik Bhargava, Bruce J. West, and Arnold J. Mandell, "Some Observations on the Question: Is Ventricular Fibrillation 'Chaos,' " preprint.
298 "IS IT POSSIBLE" Mandell.
298 "WHEN YOU REACH AN EQUILIBRIUM" Mandell.
298 MANDELL OFFERED HIS COLLEAGUES Arnold J. Mandell, "From Molecular Biological Simplification to More Realistic Central Nervous System Dynamics: An Opinion," in *Psychiatry: Psychobiological Foundations of Clinical Psychiatry* 3:2, J. O. Cavenar, et al., eds. (New York: Lippincott, 1985).
298 "THE UNDERLYING PARADIGM REMAINS" Ibid.
299 THE DYNAMICS OF SYSTEMS Huberman.
299 SUCH MODELS SEEMED TO HAVE Bernardo A. Huberman and Tad Hogg, "Phase Transitions in Artificial Intelligence Systems," preprint, Xerox Palo Alto Research Center, Palo Alto, California, 1986. Also, Tad Hogg and Bernardo A. Huberman, "Understanding Biological Computation: Reliable Learning and Recognition," *Proceedings of the National Academy of Sciences* 81 (1984), pp. 6871–75.
299 "ASTONISHING GIFT OF CONCENTRATING" Erwin Schrödinger, *What Is Life?* (Cambridge: Cambridge University Press, 1967), p. 82.
299 "IN PHYSICS WE HAVE DEALT" Ibid., p. 5.

CHAOS AND BEYOND

305 "WHEN I SAID THAT?" Ford.
305 "IN A COUPLE OF DAYS" Fox.
306 THE WORD ITSELF (Holmes) *SIAM Review* 28 (1986), p. 107; (Hao)

Chaos (Singapore: World Scentific, 1984), p. i; (Stewart) "The Geometry of Chaos," in *The Unity of Science*, Brookhaven Lecture Series, No. 209 (1984), p. 1; (Jensen) "Classical Chaos," *American Scientist* (April 1987); (Crutchfield) private communication; (Ford) "Book Reviews," *International Journal of Theoretical Physics* 25 (1986), No. 1.

306   TO HIM, THE OVERRIDING MESSAGE   Hubbard.

307   TOO NARROW A NAME   Winfree.

307   "IF YOU HAD A TURBULENT RIVER"   Huberman.

307   "LET US AGAIN LOOK"   *Gaia*, p. 125.

308   THOUGHTFUL PHYSICISTS   P. W. Atkins, *The Second Law* (New York: W. H. Freeman, 1984), p. 179. This excellent recent book is one of the few accounts of the Second Law to explore the creative power of dissipation in chaotic systems. A highly individual, philosophical view of the relationships between thermodynamics and dynamical systems is Ilya Prigogine, *Order Out of Chaos: Man's New Dialogue With Nature* (New York: Bantam, 1984).

309   GROWTH OF SUCH TIPS   Langer. The recent literature on the dynamical snowflake is voluminous. Most useful are: James S. Langer, "Instabilities and Pattern Formation," *Reviews of Modern Physics* (52) 1980, pp. 1–28; Johann Nittmann and H. Eugene Stanley, "Tip Splitting without Interfacial Tension and Dendritic Growth Patterns Arising from Molecular Anisotropy, *Nature* 321 (1986), pp. 663–68; David A. Kessler and Herbert Levine, "Pattern Selection in Fingered Growth Phenomena," to appear in *Advances in Physics*.

314   IN THE BACK OF THEIR MINDS   Gollub, Langer.

314   ODD-SHAPED TRAVELING WAVES   An interesting example of this route to the study of pattern formation is P. C. Hohenberg and M. C. Cross, "An Introduction to Pattern Formation in Nonequilibrium Systems," preprint, AT&T Bell Laboratories, Murray Hill, New Jersey.

314   IN ASTRONOMY, CHAOS EXPERTS   Wisdom; Jack Wisdom, "Meteorites May Follow a Chaotic Route to Earth," *Nature* 315 (1985), pp. 731–33, and "Chaotic Behavior and the Origin of the 3/1 Kirkwood Gap," *Icarus* 56 (1983), pp. 51–74.

314   STRUCTURES THAT REPLICATE THEMSELVES   As Farmer and Packard put it: "Adaptive behavior is an emergent property which spontaneously arises through the interaction of simple components. Whether these components are neurons, amino acids, ants, or bit strings, adaptation can only occur if the collective behavior of the whole is qualitatively different from that of the sum of the individual parts. This is precisely the definition of nonlinear." "Evolution, Games, and Learning: Models for Adaptation in Machines and Nature," introduction to conference proceedings, Center for Nonlinear Studies, Los Alamos National Laboratory, May 1985.

314   "EVOLUTION IS CHAOS"   "What Is Chaos?" p. 14.

314   "GOD PLAYS DICE"   Ford.

315   "THE PROFESSION CAN NO LONGER"   *Structure*, p. 5.

315   "BOTH EXHILARATING AND A BIT THREATENING"   William M. Schaffer, "Chaos in Ecological Systems: The Coals That Newcastle Forgot," *Trends in Ecological Systems* 1 (1986), p. 63.

315 "WHAT PASSES FOR FUNDAMENTAL" William M. Schaffer and Mark Kot, "Do Strange Attractors Govern Ecological Systems?" *BioScience* 35 (1985), p. 349.

315 SCHAFFER IS USING e.g., William M. Schaffer and Mark Kot, "Nearly One Dimensional Dynamics in an Epidemic," *Journal of Theoretical Biology* 112 (1985), pp. 403–27.

316 "MORE TO THE POINT" Schaffer.

316 YEARS LATER, SCHAFFER LIVED Schaffer; also William M. Schaffer, "A Personal Hejeira," unpublished.

Acknowledgments

MANY SCIENTISTS GENEROUSLY GUIDED, informed, and instructed me. The contributions of some will be apparent to the reader, but many others, unnamed in the text or mentioned only in passing, shared no less of their time and intelligence. They opened their files, searched their memories, debated one another, and suggested ways of thinking about science that were indispensable to me. Several read the manuscript. In researching *Chaos* I needed their patience and their honesty.

I want to express my appreciation to my editor, Daniel Frank, whose imagination, sensitivity, and integrity gave this book more than I can say. I depended on Michael Carlisle, my agent, for his exceedingly skillful and enthusiastic support. At the *New York Times*, Peter Millones and Don Erickson helped me in crucial ways. Among those who contributed to the illustrations in these pages were Heinz-Otto Peitgen, Peter Richter, James Yorke, Leo Kadanoff, Philip Marcus, Benoit Mandelbrot, Jerry Gollub, Harry Swinney, Arthur Winfree, Bruce Stewart, Fereydoon Family, Irving Epstein, Martin Glicksman, Scott Burns, James Crutchfield, John Milnor, Richard Voss, Nancy Sterngold, and Adolph Brotman. I am also grateful to my parents, Beth and Donen Gleick, who not only brought me up right but corrected the book.

Goethe wrote: "We have a right to expect from one who proposes to give us the history of any science, that he inform us of how the phenomena of which it treats were gradually known, and what was imagined, conjectured, assumed, or thought respecting them." This is a "hazardous affair," he continued, "for in such an undertaking, a writer tacitly announces at the outset that he means to place some things in light, others in shade. The author has, nevertheless, long derived pleasure from the prosecution of his task. . . . "

Index

Bold-faced page numbers indicate illustrations

351